UNDERSTANDING BATTERIES

RSC Paperbacks

RSC Paperbacks are a series of inexpensive texts suitable for teachers and students and give a clear, readable introduction to selected topics in chemistry. They should also appeal to the general chemist. For further information on all available titles contact:

Sales and Customer Care Department, Royal Society of Chemistry,
Thomas Graham House, Science Park, Milton Road, Cambridge CB4 0WF, UK
Telephone: +44 (0)1223 432360; Fax: +44 (0)1223 423429; E-mail: sales@rsc.org

Recent Titles Available

The Chemistry of Fragrances
compiled by David Pybus and Charles Sell
Polymers and the Environment
by Gerald Scott
Brewing
by Ian S. Hornsey
The Chemistry of Fireworks
by Michael S. Russell
Water (Second Edition): A Matrix of Life
by Felix Franks
The Science of Chocolate
by Stephen T. Beckett
The Science of Sugar Confectionery
by W. P. Edwards
Colour Chemistry
by R. M. Christie
Understanding Batteries
by Ronald M. Dell and David A. J. Rand

Future titles may be obtained immediately on publication by placing a standing order for RSC Paperbacks. Information on this is available from the address above.

RSC Paperbacks

UNDERSTANDING BATTERIES

RONALD M. DELL

Formerly Head of Applied Electrochemistry,
Atomic Energy Research Establishment, Harwell, UK

DAVID A. J. RAND
Group Manager, Novel Battery Technologies
CSIRO Energy Technology, Clayton South, Victoria, Australia

RS•C
ROYAL SOCIETY OF CHEMISTRY

ISBN 0-85404-605-4

A catalogue record for this book is available from the British Library

Published by The Royal Society of Chemistry,
Thomas Graham House, Science Park, Milton Road,
Cambridge CB4 0WF, UK
Registered Charity Number 207890

For further information see our web site at www.rsc.org

Typeset in Great Britain by Vision Typesetting, Manchester, UK
Printed in Great Britain by TJ International Ltd, Padstow, Cornwall

Alessandro Guiseppe Antonio Anastasio Volta (1745–1827)
Inventor of the first battery (24 March, 1800)
(with acknowledgement to Centro Volta, Italy)

Preface

Just 200 years ago, in 1800, Alessandro Volta constructed his famous 'pile' at the University of Pavia in Italy. The device was, in essence, the first 'battery' and the first source of continuous electric current. This book is written to commemorate the bicentenary of that epoch-making event, and is dedicated to the memory of Volta for laying the foundations from which modern battery science and technology have evolved.

Today, batteries are big business. Moreover, markets for batteries are likely to boom in the years ahead. Indeed, the Battelle Memorial Institute in the USA has forecast that batteries will be the second most important of ten key strategic technologies which will shape business over the next twenty years. The reasons for this are quite clear. There is a huge increase in the sales of portable electronic devices (mobile telephones, laptop and notebook computers, camcorders, *etc.*) which require battery power, and also of mains-connected devices which use batteries for memory back-up. Industry and commerce are creating a strong growth in the demand for stationary batteries to provide instantaneous power in the event of mains failure. In the transport sector, electric traction is proving more popular in environmental-sensitive areas, while major automobile manufacturers are working towards electric and hybrid electric vehicles. And, finally, the urgent need to harness so-called 'renewable energy sources' (solar energy, wind energy, *etc.*) as part of an overall strategy to ensure global energy sustainability will require more and more batteries to store the derived electrical energy and to smooth out fluctuations in both supply and demand.

There are two useful ways in which to categorize batteries. The first is to distinguish between 'primary batteries', which are discharged once and then discarded, and 'secondary batteries', which are recharged and thus used many times. The second useful distinction is between 'consumer batteries', as purchased by the individual, and 'industrial batteries' as used in industry and commerce. Most consumer batteries are small, single-cell devices (with the notable exception of the car starter battery), whereas industrial batteries tend to be much larger, multi-cell modules which are rechargeable.

A battery is a device for the storage of electrical energy in the form of chemicals and for the re-conversion of these chemicals into direct-current electricity. The processes taking place at the electrodes are quite complex, although usually they may be represented, to a first approximation, by simple electrochemical equations. During discharge, the equations define a chemical reduction process with take-up of electrons at the positive electrode, and an oxidation process with loss of electrons at the negative electrode. Recharging a secondary battery is simply a reversal of the processes which occur during discharge.

The style and contents of this book are directed towards anyone who is curious about batteries and wishes to learn more about them. In particular, it is expected that the work will be of value to engineers and technicians who are responsible for specifying, procuring and/or maintaining batteries. For the benefit of non-chemists, a deliberate attempt is made to keep the electrochemistry to as simple a level as possible. The book presents an overview of all types of battery and is by no means a comprehensive treatise on any particular type, for which specialist books and manufacturers' brochures are available.

In the opening chapter, the history of battery development is outlined and the growth of new applications and markets for batteries is discussed. This sets the scene for the remainder of the book, which is divided into three parts. Part 1 (Chapters 2 to 4) is quite general and defines the terms which are employed in battery science, shows how a battery operates, gives advice on the factors to be considered when selecting a battery for a particular application, and describes the charging of secondary batteries. The latter is the oft-neglected part of the electrochemical cycle, but is as important as the discharge conditions in determining the electrical performance and service-life of a rechargeable battery.

Part 2 is devoted to primary batteries. Zinc-based and lithium-based primary batteries of many different types are described in Chapters 5 and 6, respectively. Some specialized uses of primary batteries – in medicine, in marine environments and in defence – are discussed in Chapter 7.

Part 3 deals with rechargeable batteries and starts (Chapter 8) with the ubiquitous lead–acid battery. Batteries which employ an alkaline (potassium hydroxide) electrolyte are then examined in Chapter 9. Such batteries include the traditional nickel–cadmium battery and the more recent nickel–metal-hydride system. The latter is rapidly displacing nickel–cadmium in many applications. Rechargeable lithium batteries, especially the modern lithium-ion battery, are reviewed in Chapter 10. Over the past few years, lithium technology has made massive inroads into the portable electronics markets.

Several other secondary batteries are in various stages of development, but are not yet in general commercial production. The most significant of these so-called 'advanced batteries' are described briefly in Chapter 11. Some of the systems are available for specialist use (for instance, nickel–hydrogen batteries in satellites), while others (such as sodium–metal-chloride) are available in small quantities from pilot plants. Finally, in Chapter 12, we explore some particular areas in which rechargeable batteries are likely to play an increasingly important role. This analysis, which is analogous to that made for primary batteries in Chapter 7, again illustrates the criteria which have to be met by a battery for it to be successful in a specific application.

In summary, it is hoped that this publication will provide a better understanding of the operation and application of batteries, and will encourage readers to dig deeper into the specialist literature on the individual systems. This would be a pleasing outcome as future generations of consumers may require, at the very least, a rudimentary knowledge of battery systems given the increasing importance of these power sources as life-support systems for our rapidly evolving e-society.

The authors wish to thank all those holders of copyright who have kindly consented to the use of their illustrations. While every effort has been made, it has not always been easy to identify the correct copyright holders, especially where an illustration has appeared in more than one publication or where the holder is no longer in business. The authors apologize should any omissions have inadvertently occurred.

The authors are indebted to their friends and colleagues in the battery industry who have generously provided technical information, photographs, and diagrams. They also wish to express their special appreciation of the dedication and expert skills of Ms Rita Spiteri (CSIRO) for producing the complete text and redrafting many of the illustrations.

Contents

Abbreviations, Symbols and Units Used in Text

ABBREVIATIONS

a.c.	alternating current
AGM	absorptive glass micro-fibre
ALABC	Advanced Lead–Acid Battery Consortium (USA)
a.m.	ante meridiem (before noon)
AM	alkaline-manganese
aq.	aqueous
CCA	cold-cranking amperes
CSIRO	Commonwealth Scientific Industrial Research Organization
d.c.	direct current
DIN	Deutsche Industrie Norm (German Standard for batteries)
diss.	dissolved
DoD	depth-of-discharge
EV	electric vehicle
HEV	hybrid electric vehicle
HFEDS	US highway fuel economy driving schedule
IEC	International Electrotechnical Commission
l	liquid
LCD	liquid crystal diode
LED	light-emitting diode
LEV	low-emission vehicle
PEM	proton exchange membrane
PEO	poly(ethylene oxide)
p.m.	post meridiem (after noon)
P2VP	poly-2-vinylpyridine

RAM	rechargeable alkaline-manganese
RAPS	remote-area power supply
s	solid
SAE	Society of Automotive Engineers
SI	International System of Units
SLI	starting, lighting and ignition
SMES	superconducting magnetic energy storage
SoC	state-of-charge
SPEFC	solid polymer-electrolyte fuel cell
UDDS	US urban dynamometer driving schedule
UPS	Uninterruptible power supply
USABC	United States Advanced Battery Consortium
VRLA	valve-regulated lead–acid (battery)
ZEV	zero-emission vehicle

SYMBOLS AND UNITS – ROMAN LETTERS

A	ampere
μA	microampere
Ah	ampere-hour ($= 3600$ coulombs)
°C	degree Celsius
C_A	charge-acceptance
C_x / t	discharge rate or charge rate expressed relative to the capacity at the x hour rate (x is usually quantified); t is the discharge time in hours
cm	centimetre
dm	decimetre
e^-	electron
$E°$	standard electrode potential
g	gram
GW	gigawatt
GWh	gigawatt-hour
h	hour
Hz	hertz ($= 1$ cycle per second)
I	current
I_g	current associated with decomposition of water
J	Joule
kg	kilogram
kHz	kilohertz
km	kilometre
kPa	kilopascal
kW	kilowatt

kWh	kilowatt-hour
kWp	kilo peak-watt
l	litre
m	metre
M	molar (moles per cubic decimetre) or denotes a metal
mA	milliampere
mAh	milliampere-hour
ml	millilitre
min	minute
mm	millimetre
mΩ	milliohm
MPa	megapascal
ms	millisecond
mV	millivolt
MW	megawatt
MWh	megawatt-hour
MWp	mega peak-watt
μm	micrometre (micron)
μV	microvolt
n	number of electrons involved in an electrode process
Pa	pascal ($=9.87 \times 10^{-6}$ atmosphere)
PTC	positive temperature coefficient
R	resistance
R'	sum of the internal resistance of a cell plus the equivalent resistances of the activation and concentration overpotentials at both electrodes
s	second
S	siemens ($=$ reciprocal ohms)
t	time
T	temperature
TWh	terawatt-hour
V	volt
V	voltage
V°	standard cell voltage
V_{ch}	instantaneous cell voltage on charge
V_{d}	instantaneous cell voltage on discharge
V_{r}	reversible (open-circuit) cell voltage
W	watt
Wh	watt-hour
Wp	peak-watt
wt.%	percentage by weight
x	variable in stoichiometry

X   denotes an oxidizing agent

SYMBOLS AND UNITS – GREEK LETTERS

ε_1 efficiency of photovoltaic array
ε_2 efficiency of battery charge
ε_3 efficiency of battery discharge
η electrode overpotential
η_+ overpotential at positive electrode
η_- overpotential at negative electrode
Ω ohm
$m\Omega$ milliohm

Glossary of Battery Terms

Activation overpotential. The overpotential which results from the restrictions imposed by the kinetics of charge transfer at an electrode.

Active material. The material in the electrodes that takes part in the electrochemical reactions which store–deliver the electrical energy.

Active-material utilization. The fraction, usually expressed as a percentage, of the active material present in a positive or a negative electrode that reacts during discharge before the battery can no longer deliver the required current at a useful voltage.

Activity. A thermodynamic function used in place of concentration in equilibrium constants for reactions which involve non-ideal solutions and gases.

Ageing. Permanent loss of capacity as a result of battery use and/or the passage of time.

Alkaline battery. A battery which has a strong aqueous alkaline electrolyte.

Ampere-hour efficiency. The ratio, usually expressed as a percentage, of the ampere-hours removed from a battery during a discharge to the ampere-hours required to restore the initial capacity.

Anode. An electrode at which an oxidation process, *i.e.* loss of electrons, is occurring. (Negative electrode on discharge; positive electrode on charge.)

Aqueous battery. A battery based on an electrolyte dissolved in water.

Automotive battery. A battery designed to provide electrical power for an internal-combustion-engined vehicle.

Auxiliaries. Additional components required to operate and sustain the battery, *e.g.* heat exchangers, fans, pumps.

Battery. A multiple of electrochemical cells of the same chemistry housed in a single container. (Note, the term is often used to indicate a single cell, particularly in the case of primary systems.)

Battery conditioning. The initial application of charge–discharge cycling to establish full battery capacity.

Battery management. Regulation of both charging and discharging conditions, maintenance of battery components, and control of the operating temperature within the appropriate range.

Battery module. A battery unit manufactured as the basic component of a battery pack.

Battery pack. A number of batteries connected together to provide the required power and energy for a given application.

Bipolar. Battery design in which the component cells are connected through plates which each, in turn, act as the current-collector for the positive electrode in one cell and for the negative in the adjacent cell.

Busbar. A rigid metal connector which connects different elements within a battery. Also a rigid metallic conductor connecting battery to load.

Capacity. The amount of charge, usually expressed in ampere-hours, that can be withdrawn from a fully-charged battery under specified conditions; see also **Rated capacity**.

Cathode. An electrode at which a reduction process, *i.e.* gain of electrons, is occurring. (Positive electrode on discharge; negative electrode on charge.)

Cell. See **Electrolytic cell**.

Cell reversal. Inversion of the polarity of the electrodes of the weakest cells in a battery, usually as a result of overdischarge.

Charge. The supply of electrical energy to a battery for storage as chemical energy.

Charge-acceptance. The ability of a battery to convert active material during charge into a form which can be subsequently discharged; it is quantified as the ratio, expressed as a percentage, of the charge usefully accepted during a small increment of time to the total charge supplied during that time.

Charge profile. The sequence of current and voltage used to charge a battery.

Charge rate. The current applied to charge a battery to restore its available capacity. For a lead–acid battery, this current is usually expressed in terms of the rated capacity of the battery.

Charge retention. The ability of a battery to retain its charge under zero-current conditions.

Charge-transfer overpotential. Same as **Activation overpotential**.

Concentration overpotential. The overpotential which results from the kinetics of mass transfer of active materials to the electrode surface during the passage of current.

Couple. Combination of positive and negative electrode materials which engage in electrochemical processes in a battery.

Coulombic efficiency. Same as **Ampere-hour efficiency**.

C-rate. The discharge rate or charge rate, in amperes, and expressed relative to the rated capacity.

Current-collector. A material included in a battery to conduct electrons to or from the electrodes.

Current density. The current flowing per unit electrode area.

Cut-off voltage. The selected voltage at which charge or discharge is terminated.

Cycle. A single charge–discharge of a battery.

Cycle-life. The number of cycles that can be obtained from a battery before it fails to meet selected performance criteria.

Deep discharge. A qualitative term which indicates discharge of a large proportion of the available capacity of a battery.

Depth-of-discharge. The ratio, usually expressed as percentage, of the ampere-hours discharged from a battery at a given rate to the available capacity under the same specified conditions.

Discharge profile. The current–time sequence used in the discharge of the battery.

Discharge rate. The current at which a battery is discharged. The current can be expressed in ampere-hours, but for lead–acid batteries it is usually normalized to the rated capacity (C), and expressed as C_x/t, where x is the hour rate and t is the specified discharge time, usually in hours.

Drive-train. The elements of the propulsion system that produce and transmit mechanical energy to drive the wheels of an electric vehicle.

Duty cycle. The operating regime of a battery in terms of rate and time of both charge and discharge, and time in stand-by mode.

Efficiency. The fraction, usually expressed as a percentage, of the available output from a battery that is achieved in practice.

Electrode. An electronic conductor which acts as a source or sink of electrons which are involved in electrochemical reactions.

Electrode potential. The voltage developed by a single electrode, either positive or negative. The algebraic difference in voltage of any pair of electrodes of opposite polarity equals the cell voltage.

Electrolysis cell. An electrolytic cell in which the electrochemical reactions are caused by supplying electrical energy.

Electrolyte. The medium which provides ionic conductivity between the two electrode polarities of a cell.

Electrolytic cell. An electrochemical cell which consists of a positive and a negative electrode and an electrolyte, and which is used to store or generate electricity.

End-of-life. The stage at which a battery can no longer achieve the required performance criteria.

End voltage. Same as **Cut-off voltage**.

Energy density. The energy output from a battery per unit volume, expressed in Wh l^{-1} or W dm^{-3}.

Energy efficiency. The fraction, usually expressed as a percentage, of the energy used in charging a battery that is delivered on discharge.

Equalization. Same as **Equalizing charge**.

Equalizing charge. A charge regime designed to restore undercharged cells in a battery to the fully-charged state.

Equilibrium potential. See **Reversible potential**.

Equilibrium voltage. See **Reversible voltage**.

Expander. A compound added to the negative paste of a lead–acid battery to maintain plate porosity during charge–discharge service.

Failure mode. A process which results in a battery failing to meet the required performance criteria.

Float charging. A constant-voltage charge regime applied over extended periods to maintain a battery in the fully-charged state.

Flow battery. A battery system in which the active materials of one or both electrode polarities are stored externally and pumped to the battery during operation.

Formation. The initial charging process, during manufacture or installation, which converts the active materials into the required species for proper electrochemical operation.

Galvanic cell. An electrolytic cell in which chemical energy is converted into electrical energy on demand.

Gassing. The evolution of a gaseous product at one or both electrode polarities.

Gravimetric energy density. Same as **Specific energy**.

Gravimetric power density. Same as **Specific power**.

Grid. The framework of a battery plate that supports the active material and also serves as the current-collector.

Half-cell reaction. The electrochemical reaction at an electrode.

Hour rate. See **Discharge rate**.

Hybrid vehicle. A vehicle which operates on more than one power source.

Internal resistance. The opposition to current flow that results from the various electronic and ionic resistances within the battery.

Internal short-circuit. Same as **Short-circuit**.

IR **drop.** The decrease in battery voltage that results from current flow through the internal resistance.

Life. The duration over which a battery continues to meet the required performance criteria.

Load. When referred to battery operations, this term relates to the energy-consuming devices to which electric power is delivered.

Load voltage. The voltage of a cell or battery when delivering a current.

Maintenance-free. Description given to a battery which does not require additions of water during its normal service life.

Mass-transfer overpotential. Same as **Concentration overpotential**.

Mechanical recharging. The recharging of a battery by the physical replacement of discharged active material.

Monobloc. Same as **Battery module**.

Monopolar. The conventional method of battery construction in which the component cells are discrete and are externally connected to each other.

Name-plate capacity. Same as **Rated capacity**.

Negative electrode. The electrode in an electrolytic cell that has the lower potential.

Nominal capacity. Same as **Rated capacity**.

Nominal voltage. Same as **Open-circuit voltage**.

Ohmic loss. The decrease in the output from a battery that results from the *IR* drop.

Open-circuit voltage. The voltage of a battery when there is no net current flow.

Opportunity charging. The partial recharging of batteries, particularly traction batteries, at convenient times during the day when the batteries are not in use.

Overcharge. The supply of charge to a battery in excess of that required to return all the active materials to the fully-charged state.

Overdischarge. The discharge of a battery beyond the level specified for correct operation.

Overpotential. The shift in the potential of an electrode from its equilibrium value as a result of current flow.

Overvoltage. The shift in the voltage of a cell or battery from its equilibrium value as a result of current flow.

Parallel connection. The connection of like terminals of cells or batteries to form a system of greater capacity, but with the same voltage.

Passivation. The formation of a surface layer that impedes the electrochemical reactions at an electrode.

Paste. An intimate mixture of compounds that is applied to the grid and converted into the electrode active material.

Peak power. The sustained pulsed power which is obtainable from a battery under specified conditions, usually measured in W over 30 s.

Plate. A common term for battery electrode.

Polarity. Denotes which electrode is positive and which is negative.

Polarization. The shift in the potential of an electrode from its reversible value, or the lowering of the voltage of a cell, as a result of current flow.

Positive electrode. The electrode in an electrolytic cell that has the higher potential.

Post. Same as **Terminal**.

Power density. The power output of a battery per unit volume, usually expressed in $W\,l^{-1}$ or $W\,dm^{-3}$ and quoted at 80% depth-of-discharge.

Power pack. Same as **Battery pack**.

Primary battery. A battery designed to deliver a single discharge.

Range. The distance which an electric vehicle can travel on a single battery discharge over a specified driving profile.

Rated capacity. The capacity of a battery as specified by the particular manufacturer.

Rated power. The power capability of a battery as specified by the particular manufacturer.

Recombination. For lead–acid batteries, this term refers to the reaction of oxygen evolved on charge with the active material of the negative electrode, and with co-evolved hydrogen in the presence of a catalyst.

Redox battery. A battery in which the chemical energy is stored as dissolved redox reagents.

Reference electrode. An electrode with a reproducible, well-established potential, against which potentials of other electrodes can be measured.

Regenerative braking. The recovery of some fraction of the energy normally dissipated during braking of a vehicle and its return to a battery or some other energy-storage device.

Resistive loss. See **Ohmic loss**.

Reversible potential. The potential of an electrode when there is no net current flowing through the cell.

Reversible voltage. The difference in the reversible potentials of the two electrodes which make up the cell.

Secondary battery. A battery that is capable of repeated charging and discharging.

Self-discharge. The loss of capacity of a battery under open-circuit conditions as a result of internal chemical reactions and/or short-circuits.

Separator. An electronically non-conductive, but ion-permeable, material which prevents electrodes of opposite polarity from making contact.

Series connection. The combination of unlike terminals of cells or batteries to form a battery of greater voltage, but with the same capacity.

Service-life. The duration over which a battery continues to meet the required performance criteria for a particular application.

Shedding. The loss of active material from battery grids.

Shelf-life. The period over which a battery can be stored and still meet specified performance criteria.

Short-circuit. The direct connection of positive and negative electrodes either internal or external to the battery.

Shunt current. A self-discharge current which occurs through the electrolyte manifold in bipolar batteries because electrodes of the same polarity are at different potentials.

Specific capacity. The capacity output of a battery per unit weight, usually expressed in Ah kg^{-1}.

Specific energy. The energy output of a battery per unit weight, usually expressed in Wh kg^{-1}.

Specific power. The power output of a battery per unit weight, usually expressed in W kg^{-1}.

Standard cell voltage. The reversible voltage of an electrochemical cell with all active materials in their standard states.

Standard electrode potential. The reversible potential of an electrode with all the active materials in their standard states. Note, usual standard states specify unit activity for elements, solids, 1 M solutions, and gases at a pressure of 101.325 kPa (1 atmosphere).

State-of-charge. The fraction, usually expressed as a percentage, of the full capacity of a battery that is still available for further discharge, *i.e.* state-of-charge = [100 − (% depth-of-discharge)] %.

Storage capacity. Same as **Capacity**.

Storage-life. Same as **Shelf-life**.

String. A number of batteries connected in series.

Terminal. The external connection to a battery electrode.

Theoretical capacity. The charge output of a battery assuming 100% utilization of the active materials.

Theoretical specific energy. The energy output of a battery referred to the weight of only the active materials and a 100% utilization of these materials.

Thermal management. The means by which a battery is maintained within its operating temperature range during charging and discharging.

Thermal runaway. A condition under which an uncontrolled increase in temperature occurs and destroys the battery.

Throughput. The total energy output provided by a battery during its life, *i.e.* the sum of the energy output during each discharge.

Traction battery. A battery designed to provide motive power.

Trickle charge. A low-rate charge sufficient to compensate for self-discharge and maintain a battery in the fully-charged state.

Utilization. Same as **Active-material utilization**.

Vent. A valve which allows the release of gases from a battery but prevents spillage of electrolyte.

Voltage. The difference in potential between the two electrodes of a cell, or the two terminals of a battery.

Voltaic efficiency. The ratio, usually expressed as a percentage, of the average voltage during discharge to the average voltage during charge.

Volumetric energy density. Same as **Energy density**.

Volumetric power density. Same as **Power density**.

Watt-hour efficiency. Same as **Energy efficiency**.

Chapter 1

Energy Storage in Batteries

1.1 BATTERIES – THEIR HISTORY AND DEVELOPMENT

A battery is a chemical device for the storage of electricity. Since electricity cannot be stored directly (except in electrolytic capacitors or superconducting coils, both of which have major technical and economic limitations) it is necessary to utilize an indirect form of storage. Possibilities include the conversion of electrical energy into potential energy (pumped-hydro schemes), kinetic energy (flywheels), thermal energy (night storage heaters), or chemical energy. One form of chemical energy is hydrogen, generated by electrolysis, which may be stored and subsequently converted back into electricity in a fuel cell. Electrolysers and fuel cells are, like batteries, electrochemical energy-conversion devices. Electrolysers play an important role in the chemical industry (for instance, in the production of chlorine and caustic soda and of metals such as aluminium and copper), while fuel cells are assuming ever greater importance for localized electricity generation, combined heat and power schemes, and as power sources for electric and hybrid electric vehicles. Electrolysers and fuel cells, however, lie outside the scope of the present book.

The convenience of batteries lies in the wide range of sizes in which they may be manufactured or assembled into packs, their ability to supply electrical power instantly, their portability (for smaller sizes), and the option of single-use or multiple-use units. The last-mentioned feature provides a useful means for classifying the many different battery systems into two broad categories: (i) 'primary batteries', which utilize the chemicals once only, in a single discharge, and then are thrown away; (ii) 'secondary batteries', which may be recharged and used again. With the

1

latter batteries, the charging process involves the uptake of electricity and the conversion of the chemicals back into their original forms, so that they are available for a further discharge. Thus, battery charging is a special form of electrolytic process. The discharge–charge cycle may be repeated until the secondary battery deteriorates and its capacity to store charge falls below a practical level. Secondary batteries used to be known as 'accumulators' but the term has largely dropped out of use, at least in the English language. It is quite common (and acceptable) for secondary batteries to be referred to as 'rechargeable batteries'.

It is just 200 years since the invention of the first battery. This has been ascribed to Alessandro Volta (1745–1827), Professor of Natural Philosophy (physics) at the University of Pavia in Italy. His name is commemorated for all time by the unit of electrical potential, the volt. Volta's famous experiment, described in a letter to the Royal Society of London in 1800, consisted of assembling a pile of alternate silver (or brass or copper) and zinc (or tin) discs, with each pair of dissimilar metals separated from the next by a piece of cloth which was saturated with brine. One end of the pile terminated in a silver disc and the other in a zinc disc, and a continuous current of electricity was produced as soon as the two were connected by a wire conductor. This was the first galvanic, or primary, battery and became known as 'Volta's pile'. Batteries have come a long way in 200 years! It is interesting to note that the French word for battery ('la pile') stems directly from Volta's device.

The next significant step in the development of batteries was the invention of the Daniell cell by John Daniell (1790–1845), Professor of Chemistry at King's College, London. In 1836, Daniell took a copper vessel filled with copper sulfate solution and in it immersed a gullet of an ox. This unusual, and somewhat repugnant, receptacle contained a solution of sulfuric acid and a vertical zinc rod. Discharge of the resulting cell caused the zinc electrode to dissolve and copper to be deposited at the positive electrode. The cell gave a voltage of 1.1 V. This was possibly the first practical galvanic cell to give a continuous current of useful magnitude. Further modifications (Figure 1.1) included the use of porous ceramic pots ('separators') instead of animal membranes, substitution of sulfuric acid by zinc or magnesium sulfate, and the development of multi-cell batteries. Daniell cells were adopted by commercial telegraphic systems following a rapid expansion of such services in the early 1850s.

A subsequent major advance was made by the French chemist Georges Leclanché (1839–1882) who, in 1866, invented the primary cell which bears his name (see Section 5.1, Chapter 5). This consists of a zinc rod as the negative electrode and a carbon rod as the positive electrode, both immersed in a solution of ammonium chloride contained in a glass jar.

(a)

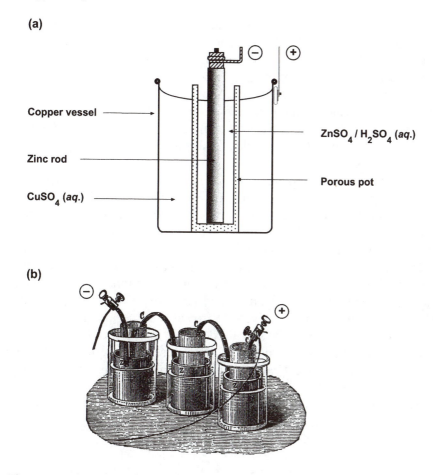

Copper vessel

Zinc rod

$CuSO_4$ (aq.)

$ZnSO_4 / H_2SO_4$ (aq.)

Porous pot

(b)

Figure 1.1   (a) *Construction of a Daniell cell*; (b) *early illustration of Daniell cells (reverse configuration with Z = zinc, C = copper)*

The positive electrode was housed in an inner porous ceramic pot and packed around with a mixture of powdered manganese dioxide and carbon. The cell, which has been extensively developed ever since, gives a voltage of 1.5 V. A major advance took place in the late 19th century when the idea of using a zinc can as both container and electrode was patented and came into general use.

Before the invention of these galvanic cells, the only electricity known and available was static electricity, as produced by friction between dissimilar materials or in thunderstorms. The first recorded electrostatic machine was made in 1663 by the German physicist Otto von Guericke (1602–1686). In 1797, George Pearson (1751–1828) reported that in order to electrolyse water the electrostatic machine had to be discharged 14 600 times, and then produced only about 5.5 ml of a gaseous mixture of hydrogen and oxygen. Not surprisingly, therefore, the availability of a continuous current from a galvanic cell caused a revolution in technology. It formed the basis of telegraphy and of the electric door bell, and later of radio reception. In the chemical field, the supply of electricity allowed the development of electroplating, electroforming and, for the first time, the electrolytic extraction of metals such as sodium, potassium, and calcium. Copper, for use in electrical equipment, could be purified by electrorefining.

The first effective demonstration of a secondary (rechargeable) cell was given in 1859 by the French chemist Gaston Planté (1834–1889). This cell consisted of two concentric spirals of lead sheet, separated by porous cloth and immersed in dilute sulfuric acid within a cylindrical glass vessel [Figure 1.2(a)]. The 'lead–acid battery' gave an output of 2 V, but very little current initially because of the low surface area of the plates. By a series of discharges and charges, the chemical reactions at the surface of the plates resulted in the gradual build-up of deposits of higher surface area and the current progressively improved. This became known as the 'formation process', a term still used today in the initial charging of lead–acid batteries. In March 1860, Planté presented a battery of ten cells (20 V) to the Académie Française in Paris; an illustration of an early battery of Planté cells is shown in Figure 1.2(b).

An important advance in the technology of the lead–acid battery was

(a)   **(b)**

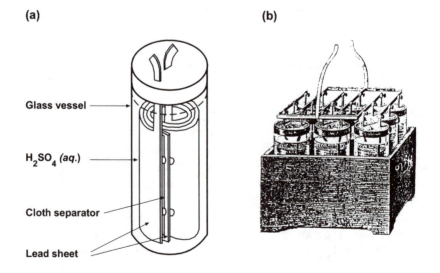

Glass vessel

H$_2$SO$_4$ (aq.)

Cloth separator

Lead sheet

Figure 1.2   (a) *Schematic of Planté's lead–acid cell*; (b) *early illustration of a battery of nine Planté cells.*

achieved by the French chemical engineer Camille Fauré (1840–1898) who, in 1881, showed how the level of electrical charge, or the 'capacity', of the system could be greatly increased by coating the lead plates with a paste of lead dioxide and sulfuric acid. This process also reduced the time of plate formation from months to hours, and thus became part of the basic technology of the lead–acid battery industry. Without the ingenuity of Planté and Fauré, car engines would still be started by hand cranking!

Around the turn of the 19th century, the first work on batteries with alkaline solutions was reported. In Sweden, Waldemar Jungner (1869–1924) took out a patent for a 'nickel–cadmium battery'. This had a positive electrode of nickel hydroxide and, as the negative, a mixture of cadmium and iron powders. The electrodes were immersed in a concentrated solution of potassium hydroxide. At about the same time, Thomas Edison (1847–1931), who was working independently in the USA, developed and patented the 'nickel–iron battery'. This system was similar in principle, except that it used an all-iron negative electrode rather than a cadmium–iron mixture. Both cells gave 1.2 V. There was considerable rivalry between the two inventors when they learnt of each other's work and patent suits ensued. Ultimately, both alkaline batteries were brought into commercial use (see Sections 9.3 and 9.4, Chapter 9). These, in outline, were the principal discoveries of the 19th century upon which the modern battery industry is founded.

The 20th century has seen major advances in battery science and technology. Primary zinc (Leclanché) batteries have been greatly improved by the invention of alkaline batteries (Chapter 5), while advances in materials technology and cell design have revolutionized the performance of the lead–acid battery (Chapter 8). Entirely new secondary batteries have been developed and commercialized, notably the nickel–metal-hydride battery (Chapter 9) and the lithium-ion battery (Chapter 10). Many batteries for specialized applications are now manufactured, and these are introduced throughout this book. Finally, there are a number of secondary batteries in an advanced stage of development, but not yet fully commercialized (Chapter 11). Certainly, at the beginning of a new millennium, battery science and development is flourishing.

1.2 BATTERY APPLICATIONS AND MARKETS

In the 1940s, the principal domestic uses for batteries were in pocket torches (flashlamps), in a few toys, in vehicles (for starting, lighting and ignition), and in radios. Before the advent of transistors and solid-state electronics, when thermionic valves (tubes) were used in radios, it was necessary to have a source of low-voltage electricity to heat the filament and a d.c. high voltage to accelerate electrons between the cathode and the anode of the valve. Many radios in those days were remote from the electricity supply and in such circumstances two batteries were required, namely, a 2 V lead–acid 'accumulator' to supply the filament current and a 'high tension' battery consisting of 100 or 120 Leclanché cells wired in series to supply the high d.c. voltage. The radio weighed several kilograms, had to be maintained in an upright position to avoid spillage of acid, and was anything but 'portable'. No doubt there were other applications for batteries in industry and commerce, for example emergency lighting and back-up for the primitive telephone systems of the day.

Over the past 50 years, the applications for small sealed batteries in the home (consumer batteries) have expanded phenomenally. Today, small primary or rechargeable batteries are employed in a huge number of appliances. Some examples are as follows.

- *Household*: telephones, clock-radios, security alarms, smoke detectors, portable fluorescent lamps, torches and lanterns, door-chimes, car central-locking activators.
- *Workshop and garden*: portable tools (*e.g.*, screwdrivers, drills, sanders), portable test meters, hedge trimmers, lawnmowers.
- *Entertainment*: portable radios and televisions, compact disc players, tape recorders, remote controllers for televisions and videos, electronic games and toys, keyboards.
- *Personal hygiene and health*: toothbrushes, bathroom scales, hair trimmers, shavers, blood-pressure monitors, hearing aids, heart pacemakers.
- *Portable electronic devices*: watches and clocks, cameras, camcorders, calculators, organizers, mobile telephones, laptop and notebook computers, bar-code readers.

The average family may now have as many as 40 to 60 consumer batteries at any one time in and around the home. Many of these will be batteries of advanced design and construction that give greatly improved performance as a result of developments in materials science and technology. Although most small consumer batteries are still of the primary ('throw-away') variety, there is a growing trend to adopt rechargeable batteries as being more economical.

Moving to larger secondary batteries, principally lead–acid, there has been a growth in the market for engine-starter batteries to mirror the growth in transport. Every internal combustion engine – whether it be in a car, a van or truck, a bus, a ship, an aircraft, or even a static engine – requires a starter battery. In the road transport market, these used to be known as 'starting, lighting, ignition (SLI) batteries' to mirror their principal functions. With vehicles becoming ever more sophisticated, the required number of electric motors and other electrical facilities has mushroomed. The demands upon the battery have grown correspondingly, to the point that soon it may be necessary to fit two batteries to the private car; one battery for engine starting, the other for running the auxiliaries (see Section 8.3.5, Chapter 8). These developments have resulted in a change in name from 'SLI batteries' to 'automotive batteries' to reflect the many other functions of the modern car battery.

There has been a similar proliferation in the demand for very large, installed battery packs. Almost every public building (airports, hospitals, hotels, railway stations, stores and supermarkets, *etc.*) must have an uninterruptible power supply (UPS) and this requires a battery pack to take over seamlessly when the mains supply fails, until such time as a local generator can be started. Most aircraft, factories, office blocks, ships, telephone exchanges and even power stations have their own uninterruptible supply. Other applications for large primary or rechargeable batteries are in the defence field (missiles, submarine traction batteries, torpedoes), in space vehicles (satellites, space probes), in solar energy storage (cathodic protection systems, electric fences, navigation beacons, remote-area power supplies, telecommunications, water pumping), and in electric and hybrid electric vehicles (bicycles, fork-lift trucks, golf-carts, invalid conveyances, specialist road vehicles, tugs and tractors, a few cars). Although many of these applications utilize long-established lead–acid or nickel–cadmium rechargeable batteries, these have been transformed into superior units in recent years by major advances in materials and design. In the past decade, new rechargeable batteries have been introduced into consumer markets. Moreover, there are now specialized batteries with unconventional chemistries for use in the military and the space fields.

The size range of batteries is now enormous. The scale extends from small button cells of energy content ~ 0.1 Wh, all the way to load-levelling batteries of energy content ~ 10 MWh. (Note, a battery of energy content 0.1 Wh is theoretically capable of supplying a power of 0.1 watt for an hour, or 1.0 watt for a tenth of an hour, although in practice the recoverable energy does depend upon the discharge rate.) Few other commodities come in such a range of sizes, certainly not pots of paint or packets of cornflakes! Some typical applications for batteries of differing size are given in Table 1.1.

The size and value of the market for different classes of battery is difficult to estimate. The mass markets are in consumer primary and secondary cells, automotive batteries, and in industrial batteries for

Table 1.1 *Battery sizes and applications*

Battery type	Stored energy (Wh)	Applications
Miniature/button cells	0.1–0.5	Watches, calculators
Portable communications	2–100	Mobile telephones, laptops
Domestic uses	2–100	Portable radios and televisions, flashlamps, toys, camcorders, power tools
Automotive	10^2–10^3	Starter batteries for cars, trucks, buses, boats, *etc.* Traction for golf carts and invalid wheelchairs
Remote-area power supplies	10^3–10^5	Household/village electricity supplies, lighting, telecommunications, water pumping
Traction	10^4–10^6	Electric vehicles, fork-lift trucks, tractors
Stationary	10^4–10^6	Stand-by batteries (UPS)
Submarine	10^6–10^7	Underwater propulsion
Utility energy-storage	10^7	Load-levelling, peak-shaving, spinning reserve, voltage regulation, end-of-line back-up

stationary and mobile power. As long ago as 1989, the annual markets in the USA for primary and secondary batteries were estimated as US$2.3 billion and US$4.2 billion, respectively. In 1991, the world battery market was said to be US$21 billion; 40% by value was supplied by primary batteries and 60% by secondary batteries. Undoubtedly, the market has grown since then. A more recent (1997) estimate gave the world demand for small, consumer secondary cells as a little over two billion units, of value US$3–5 billion. Add to this the huge demand for primary cells and larger rechargeable batteries and it is clear that this is indeed a substantial industry. Moreover, the major expansion of the battery industry in developing nations, such as China, and the present trend towards more sophisticated, higher cost batteries (such as alkaline-manganese primary batteries, and nickel–metal-hydride and lithium-ion secondary batteries) are greatly accelerating the industry turnover. For example, global sales of nickel–metal-hydride batteries have grown from 310 to 800 million units between 1995 and 1998, while sales of lithium-ion batteries have increased from 35 to 280 million units. Today, the battery industry world-wide is a vast business with strong prospects for substantial growth in the near future.

Part 1

Know Your Battery

Before introducing individual types of primary and secondary battery – in Parts 2 and 3, respectively – we consider, in Part 1, some of the more general themes which are common to all batteries.

Chapter 2 describes how a battery operates and introduces the technical terms which are used in battery science and technology. A knowledge of these terms is needed to understand the performance characteristics of individual batteries. Different batteries utilize different chemical reactions at the positive and negative electrodes, and it is shown how these reactions give rise to a cell voltage. Some simple discharge and charge curves, *i.e.* cell voltage plotted against time, are presented for several battery chemistries.

In Chapter 3, there is a discussion of the many factors which play a role in determining the choice of battery to be made for a given application. It is emphasized that a detailed consideration should first be made of the duty and service-life demanded from the battery. These requirements must then be matched with the known performance of different battery types, with regard also to further constraints such as the availability, cost and environmental acceptability of the battery materials themselves. Often a compromise has to be accepted since rarely does a battery type meet all the required specifications.

The charging of secondary batteries is discussed in Chapter 4. It should be noted that, to achieve good performance and service-life, the conditions under which a battery is charged are equally as important as the accompanying discharge conditions. Various charging regimes are described, together with the types of charger that are commercially available and the procedures available for effective charging of industrial batteries and electric road vehicle batteries. Finally, a description is given of equipment which is used for the monitoring, management and control of charge–discharge operations to ensure that batteries attain their design lives.

Chapter 2

How a Battery Operates

In this chapter we define the terminology used in battery science and describe, in simple terms, how a battery operates. We begin with some definitions.

An **electrolytic cell** (more commonly called an **electrochemical cell**) is a chemical device for generating electricity. It consists of a **positive electrode** and a **negative electrode**, separated by an **electrolyte** which is capable of conducting ions between the two electrodes, but which is itself an electronic insulator. If the electrolyte were an electronic conductor it would give rise to **self-discharge**, and/or an **internal short-circuit** would develop within the cell. The majority of battery electrolytes are concentrated aqueous solutions of acids, alkalis, or salts. Other ionic conductors which have been used as electrolytes are organic salt solutions, polymers, ceramics, and fused salts.

When two or more cells are joined together electrically, either in series (connecting positive electrode to negative electrode) or, less usually, in a series–parallel array, the resulting assembly is termed a **battery**. Strictly speaking, a battery is a multi-cell array, although in common usage, many single cells, particularly primary cells, are called 'batteries'; we shall follow this practice and use the term 'battery' to encompass both single and multi-cell devices housed in a single container.

Larger sizes of rechargeable battery have further complications of nomenclature. It is common for a single cell to have a number of **electrode plates** of each polarity joined together in parallel (positive plate to positive plate, negative plate to negative plate) and immersed in the same electrolyte. The parallel plates of each polarity are interleaved. This arrangement increases the **storage capacity** of the cell (measured in coulombs or, more practically, ampere-hours) while keeping size to a minimum but, as the plates are connected in parallel, the voltage is still that of a single cell. In the common car battery, for example, six such

multi-plate cells, each giving 2 V, are series-connected in a single enclosure to produce a 12 V battery. Such a unit is also referred to as a **battery
module** or **monobloc**. An assembly of monoblocs, as for example the
stand-by battery in a telephone exchange or the traction battery in an
electric vehicle, is termed a **battery pack** or **power pack**.

2.1   CHEMISTRY IN CELLS

How does a chemical device store and generate electricity? An electrolytic
cell is shown schematically in Figure 2.1(a). The essential components are
a positive and negative electrode, an electrolyte, a separator and a housing (or container). The positive and negative electrodes have to be as close
to each other as possible (not as presented in the schematic diagram), so
as to minimize the **internal resistance** of the cell. Typically, this resistance
is of the order of milliohms (mΩ), so that the voltage drop across the
battery terminals shall not be too great when drawing heavy currents.
Even with the best ion-conducting electrolytes, this low resistance is

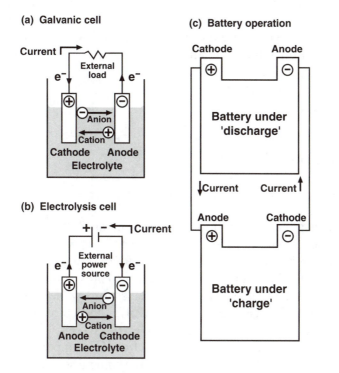

Figure 2.1   *Schematic representation of the operation of electrochemical cells and batteries*
(By courtesy of Research Studies Press Ltd)

possible only when the separation between the electrodes is restricted to about 1 mm. The **separator** is a thin, usually porous, insulating material (not shown in Figure 2.1) whose role is to prevent the two electrodes from touching each other and, thereby, short-circuiting. Without the separator, this would inevitably occur, either on cell assembly or during extended operation. The pores of the separator are filled with electrolyte and the ionic current is conveyed through these pores. When there is excess of liquid electrolyte, we use the term **flooded-electrolyte battery**; but if the liquid electrolyte is totally immobilized within the separator we refer to a **starved-electrolyte battery**.

The chemical reactions which generate electricity take place at the two electrodes. Each electrode undergoes a **half-cell reaction**. An electrode is made up of the chemicals which undergo reaction, known as the **active material** (or **active mass**), and this is attached to a metal component, the **current-collector** (or 'grid'). The driving force for the external current derived from a cell is the difference in the **electrode potentials** of the two half-cell reactions. An electrolytic cell which produces a current is a **galvanic cell**. During discharge of the cell, the current-collector of the negative electrode gathers the electrons liberated by the chemical reaction. These electrons pass through the external load, thereby doing useful work, and are accepted by the current-collector of the positive electrode for participation in the complementary reaction. Thus, we may represent the chemistry as follows.

At the negative electrode:

$$M \rightarrow M^{n+} + ne^-$$

(2.1)

At the positive electrode:

$$nX + ne^- \rightarrow nX^-$$

(2.2)

where: M is a metal; X is an oxidizing agent, such as a metal oxide in a high valence (oxidation) state; e^- is an electron.

During discharge of a battery, negative ions (anions) move towards the negative electrode, and positive ions (cations) move towards the positive electrode. The flows of these ions are reversed during the charging process [Figure 2.1(b)]. The charging of a cell involves an electrolysis process and so, during charge, the cell becomes an **electrolysis cell**. Two cells connected in opposition are shown in Figure 2.1(c). If the upper cell is of higher voltage than the lower cell, then it is undergoing discharge and is charging the lower cell.

During discharge, the reaction at the negative electrode is an **oxidation** (or **anodic) reaction**, with liberation of electrons, and that at the positive electrode is a **reduction** (or **cathodic) reaction**, with uptake of electrons. Since batteries are generally considered to operate in the discharge mode, the negative electrode is often known as the **anode** and the positive electrode as the **cathode**, *i.e.* the exact opposite of the convention for electrolysis. Obviously, during recharging of a secondary battery, the anode becomes the cathode and the cathode becomes the anode (Figure 2.1). Nevertheless, the negative electrode remains a negative electrode, and the positive electrode remains a positive electrode. Thus, to prevent confusion, it is better to avoid the use of the terms anode and cathode and to adhere to negative and positive electrodes.

Typical metals which form the negative active-mass are zinc (Zn), cadmium (Cd), lead (Pb) and lithium (Li). The positive active-mass is likely to be an oxide of manganese (MnO_2), nickel (NiOOH) or lead (PbO_2), in a higher valence state; or sometimes a metal sulfide or halide; or even elemental oxygen, sulfur, or a halogen such as bromine (Br_2). [Note, the 'valence' of an ion is equal to its charge, *e.g.* in KOH, potassium (K^+) has a valence of one, signified as K(I).] These different possibilities for the active materials are referred to as different **cell chemistries**. Where the electronic conductivity of the positive active-mass is inadequate to convey the electrons to the actual reaction sites, it is conventional to mix the active material with a conducting substance, such as carbon black or graphite, and so provide an improved pathway for the electrons.

One of the best known rechargeable batteries is the **lead–acid battery**, familiar to all as the car starter (automotive) battery. This has lead dioxide (PbO_2) as the positive active-material, lead (Pb) as the negative active-material, and sulfuric acid (H_2SO_4) as the electrolyte. The two half-cell reactions are as follows.

At the positive electrode:

$$PbO_2 + HSO_4^- + 3H^+ + 2e^- \underset{\text{Charge}}{\overset{\text{Discharge}}{\rightleftarrows}} PbSO_4 + 2H_2O \quad E° = +1.690 \text{ V} \quad (2.3)$$

At the negative electrode:

$$Pb + HSO_4^- \underset{\text{Charge}}{\overset{\text{Discharge}}{\rightleftarrows}} PbSO_4 + H^+ + 2e^- \quad E° = -0.358 \text{ V} \quad (2.4)$$

where $E°$ is the standard electrode potential for each cell reaction (see below).

The overall cell reaction is:

Discharge

$$Pb + PbO_2 + 2H_2SO_4 \; \underset{\text{Charge}}{\overset{\longrightarrow}{\longleftarrow}} \; 2PbSO_4 + 2H_2O \quad V^\circ = +2.048 \text{ V} \qquad (2.5)$$

where V° is the standard cell voltage (see below). It may be noted that the lead–acid battery is unusual among batteries in that the electrolyte takes part in the half-cell reactions. Generally, this is not the case and the electrolyte serves only to conduct ions from one electrode to the other.

The voltage of a cell is the difference in the potentials of the two component electrodes. The absolute potential of any given electrode cannot be determined since all practical methods of measurement of electrode potential depend on completing the electrical circuit with a second electrode and this, too, will exert a potential. It is possible, however, to obtain a relative value by coupling the electrode to a second, arbitrarily chosen, reference system. By convention, the potential of the hydrogen electrode under **standard conditions** (in practice: $H_2SO_4 \sim 0.5$ M; H_2 pressure = 101.3 kPa; temperature = 25 °C) is set at zero volts, and all electrode potentials are referred to this **standard hydrogen electrode**. When the electrode under measurement is also in a standard state, the **standard electrode potential** (E°) is obtained. The signs of electrode potentials are derived on the basis of reduction processes. Thus, the positive electrode of a cell, which undergoes a reduction reaction on discharge [*e.g.* for lead–acid, Equation (2.3)], has a positive potential; whereas the negative electrode, which experiences an oxidation reaction [*e.g.* Equation (2.4)], has a negative potential.

The difference between the potentials of the positive and negative electrodes gives the **reversible voltage** (V_r) or **open-circuit voltage** of the cell, that is the voltage across the terminals of the cell when there is no net current flow. Under standard conditions, this is the **standard cell voltage** (V°). Thus, for the lead–acid battery: $V^\circ = 1.690 - (-0.358)$ V $= 2.048$ V and the cell reaction proceeds from left to right, as represented by Equation (2.5).

2.2   CELL DISCHARGE AND CHARGE

The voltage of a battery measured **on load** (*i.e.* when drawing current) will be lower than that on open-circuit. This results from the internal impedance of the battery which is made up of:

- **polarization losses** at the electrodes;
- **ohmic (resistive) losses** [*i.e.* current (I) × resistance (R) $= IR$] in the current-collectors, electrolyte and active masses.

When current flows through a battery, there is departure from equilibrium conditions and the useful work performed by the battery is less than the maximum value. The shift in potential of an electrode away from the reversible (equilibrium) value is termed the **electrode overpotential (η)**. This overpotential is made up of two components:

(i) an **activation overpotential** caused by kinetic limitations to the charge-transfer process at the electrode; this is an intrinsic property of the electrode material immersed in the electrolyte, *i.e.* an interface phenomenon.

(ii) a **concentration overpotential** which results from depletion of reactants in the vicinity of the electrode due to slow diffusion from the bulk solution or across the product layer; this is an extensive property that is dependent upon the thickness and porosity of the electrode and the ease of diffusion through it, as well as upon mass-transport processes in the electrolyte.

Together, these two overpotentials result in a voltage drop at the electrode, which is termed the polarization loss, and the electrode is said to be **polarized**. Similarly, the voltage drop due to the internal resistance of the battery (*i.e.* resistive losses) is commonly referred to as **resistance** (or **ohmic) polarization** (or **overpotential**). A substance which reduces electrode polarization, *e.g.* manganese dioxide in alkaline primary cells, is sometimes known as a **depolarizer**.

Electrodes employed in rechargeable batteries undergo reversible discharge/charge reactions. An example is that of the nickel oxyhydroxide positive electrode which is used in rechargeable alkaline-electrolyte batteries. The reversible reaction of this electrode is:

$$\text{NiOOH} + \text{H}_2\text{O} + \text{e}^- \underset{\text{Charge}}{\overset{\text{Discharge}}{\rightleftharpoons}} \text{Ni(OH)}_2 + \text{OH}^- \tag{2.6}$$

where trivalent nickel, Ni(III),* is being reduced on discharge to divalent nickel, Ni(II). For this process to take place, an electron from the current-collector has to interact with a water molecule and a particle of the solid nickel(III) oxyhydroxide. On a macroscopic scale, this will begin with particles of solid immediately adjacent to the current-collector as these

*NiOOH is an oxy-hydroxide of trivalent nickel and not a peroxide.

are encountered first by the emerging electron. As discharge proceeds, a reaction front will propagate into the bulk of the active material and away from the current-collector. On a microscopic scale, and superimposed upon the macroscopic reaction front, the individual particles of solid reactant will become coated with the reaction product, $Ni(OH)_2$, as shown schematically in Figure 2.2. Further conversion (discharge) of each particle will depend upon the diffusion of an electron and a water molecule through the layer of reaction product and into the core of the particle. Most batteries perform better if allowed to 'rest' or 'recuperate' periodically during discharge. This permits diffusion to take place that reduces concentration gradients in the active mass and restores a balanced situation, nearer to equilibrium. On charging the battery, the reverse processes take place, again with diffusion controlling both the macroscopic and the microscopic reaction paths within the active mass.

Polarization losses occur at each electrode and are responsible for a decreased cell voltage during discharge (V_d) and an increased cell voltage on charge (V_{ch}), i.e.

$$V_d = V_r - \eta_+ - \eta_- - IR \tag{2.7}$$

$$V_{ch} = V_r + \eta_+ + \eta_- + IR \tag{2.8}$$

where η_+ and η_- are the overpotentials at the positive and negative electrodes, respectively. At low overpotentials, the relationship between η

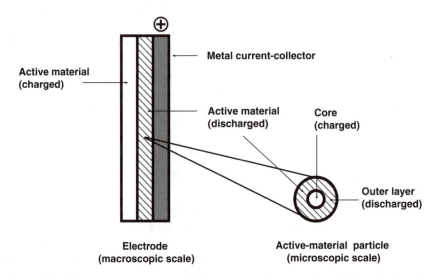

Figure 2.2 *Schematic representation of the reaction front progressing through the active material of a positive electrode undergoing discharge*

and I takes the same form as Ohm's Law and, under these conditions, Equations (2.7) and (2.8) reduce to:

$$V_d = V_r - IR' \qquad (2.9)$$

and

$$V_{ch} = V_r + IR' \qquad (2.10)$$

where R' is the sum of the internal resistances of the cell and the equivalent resistances of the activation and concentration overpotentials at both electrodes. The relationship between the practical cell voltage, on both discharge and charge, and the reversible cell voltage is presented schematically in Figure 2.3. To illustrate how this works in practice, Figure 2.4 shows some experimental curves for the discharge and charge of a small sodium–metal-chloride cell (see Section 11.6, Chapter 11). As is immediately apparent, the practical discharge voltage lies below the reversible cell voltage of 2.35 V, and the practical charge voltage lies above this value. The deviation from V_r is a measure of the combined influence of the internal resistance of the cell and the polarization losses. As the discharge (or charge) current increases, the deviation becomes greater, in accordance with Equations (2.9) and (2.10).

If a cell/battery is discharged at constant current, the storage capacity

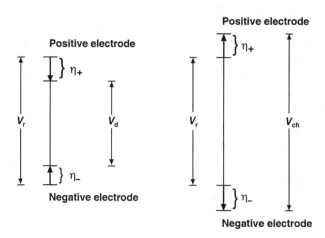

(a) Discharge   **(b) Charge**

Figure 2.3   *Schematic representation of the relationship between practical cell voltages and reversible cell voltage*
(By courtesy of Research Studies Press Ltd)

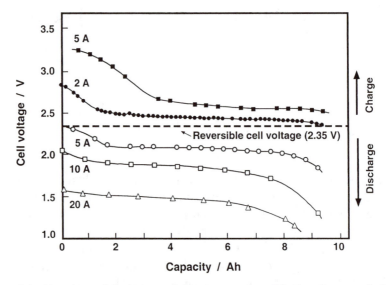

Figure 2.4 *Experimental discharge and charge curves for a 7.7 Ah sodium–metal-chloride cell at various currents (open symbols represent discharge; closed symbols represent charge)*

is the product, expressed in **ampere-hours**, of the current and the number of hours for which the cell/battery can be discharged to a defined cut-off voltage. If the current is not constant, as is often the case, then it is the integral of the current over the time of discharge. The ampere-hour is a practical unit and is expressed in SI units as: 1 ampere-hour = 3600 coulombs. The value of the capacity depends not only upon the ambient temperature and upon the age/history of the cell, but also, to a greater or lesser extent, upon the rate of discharge employed. The higher the rate of discharge, the less the available capacity. This is particularly so for lead–acid batteries, as shown in Figure 2.5 for a battery of nominal capacity 100 Ah when discharged at the 5-h rate. It will be at once apparent that the capacity delivered at the 30-min rate to a cut-off of 1.7 V (30 Ah) is less than one-third of that at the 5-h rate, while slow discharge at the 10-h rate gives an extra 20% capacity over the nominal value. At the same time, as we have seen, the operating voltage falls off markedly at high discharge rates because of polarization. The result is that the **stored energy delivered**, measured in **watt-hours** (Wh = V × Ah), declines even more sharply at higher discharge rates. The watt-hour, like the ampere-hour, is a practical unit and is expressed in SI units as 1 Wh = 3600 Joules.

Considering the capacity and the stored energy available from a battery, it is essential to define both the discharge rate and the temperature to be employed. Battery manufacturers generally state a **rated capacity**

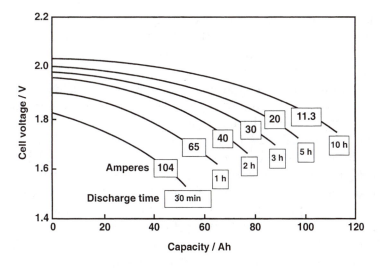

Figure 2.5 *Discharge curves for a lead–acid battery of nominal capacity 100 Ah at the 5-h rate*
(By courtesy of Research Studies Press Ltd)

(so-called **nominal** or **name-plate capacity**) under specified discharge conditions, often the 5-h rate at 25°C to a designated cut-off voltage. For convenience, the term C/t (where t is the discharge time in hours) is often used to define the rate at which the capacity is calculated. Thus, $C/5$ is the 5-h discharge rate and $2C$ is the 30-min rate. A refinement of this, sometimes used for rechargeable batteries, is to express the discharge rate as C_x/t where x is the hour rate at which the capacity is defined. For example, $C_5/5$ is the 5-h discharge rate of a battery whose capacity measured at the 5-h rate is C_5. Thus, the $C_5/5$ rate is 20 A for a battery of nominal capacity 100 Ah at the 5-h rate. Discharge of the same battery at 50 A would be stated as the $C_5/2$ rate. The latter is simply an expression of the current withdrawn and does not signify that the battery can be discharged at this current for 2 h. As shown in Figure 2.5, if the battery is to operate effectively for 2 h, then it must be discharged at a current of no more than 40 A, *i.e.* the $C_2/2$ rate. To ensure that batteries will meet their specification, allowing for variability in manufacture and use, actual capacities are often engineered to be greater than the nominal values.

The amount of charge (capacity) withdrawn compared with the total amount which is available at the same discharge rate is the **depth-of-discharge** (DoD). This ratio is usually expressed as a percentage. It follows that the **state-of-charge** (SoC) of a battery is the fraction of the full capacity that is still available for further discharge, *i.e.* SoC = $[100 - (\%$ DoD)$]\%$.

Another term which is sometimes employed, particularly when considering individual electrodes rather than cells or batteries, is the **current density**. This is the current per unit area of electrode, generally expressed as milliamperes per cm^2 ($mA\ cm^{-2}$).

The cut-off voltage of a battery discharge, used to define its capacity, can be a somewhat arbitrary figure. Some batteries, for instance the cadmium–nickel-oxide battery (generally referred to as 'nickel–cadmium'), have quite flat discharge curves, from 80 to 20% of their capacity, particularly at low current drains, with an abrupt fall-off in voltage towards the end of discharge (Figure 2.6). It is then relatively easy to define the end-point of discharge. For many other batteries, the discharge voltage declines progressively and the end-point is fairly arbitrary. Such is the case with the lead–acid battery (Figure 2.5), for which an average of 1.7 V per cell is often taken as the cut-off limit.

When dealing with small consumer batteries, discharge at constant current is not a particularly useful concept as currents are not usually measured and, anyway, decline as the cell is discharged. A more practical measure is the service-life, in hours, as a function of the resistive load imposed by the equipment. For example, Figure 2.7 shows the discharge curves for an alkaline-manganese (AM) AA-size cell (see Chapter 5) as a function of the applied load, to a cut-off voltage of 0.8 V. The service-life at 21 °C ranges from 140 h for a 62 Ω load, down to 8 h for a 3.9 Ω load. Information such as this for varying cell sizes is particularly useful to the design engineer who wishes to optimize the combined battery–application system.

This chapter has provided an overview of how a battery operates and

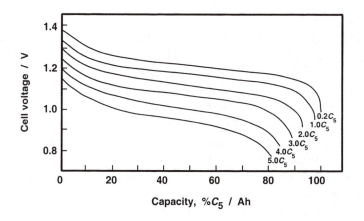

Figure 2.6   *Discharge curves for a high-power, nickel–cadmium traction battery (SAFT electric-vehicle battery) at 20°C.*
(By courtesy of SAFT)

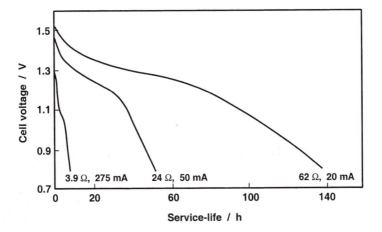

Figure 2.7 *Discharge curves for a AA-size alkaline-manganese primary cell*
(By courtesy of Duracell Batteries)

has introduced some of the principal terms which are employed in battery science. The concepts of stored energy and of power output, particularly in relation to battery mass and volume, are discussed in the next chapter.

Chapter 3

How to Choose a Battery

The choice of a battery for an intended application is dictated by considerations of performance and cost. These are not necessarily independent of each other. As with other merchandise, it is possible to purchase a quality product at a premium price or an inferior, but adequate, product at a lower price. Cost can be a primary consideration when, for example, persons of limited means buy replacement batteries for flashlamps or for their cars, but a secondary consideration when performance and reliability are vital, as in many safety applications. One example where performance and reliability are dominant is for the battery in a communications satellite, where battery failure means loss of the entire satellite as there is no practical means of effecting a replacement.

Consumer batteries are often sold on price, although it is quite common for battery advertising or attractive labelling to induce consumers to buy quality products for purposes where cheaper batteries would be perfectly satisfactory. Industrial batteries, on the other hand, are usually purchased by engineers who are more likely to study in detail the manufacturers' brochures in order to select, on the grounds of both performance and cost, the best unit for the intended use. In this chapter, we describe the various technical parameters which are important for battery selection and, by means of examples, show how these match the requirements of different applications.

3.1 PRIMARY OR SECONDARY?

The first decision to be made is whether the application merits the purchase of a primary or a secondary battery. In many instances, the choice is self-evident. Clearly, nobody would purchase a rechargeable button cell for a watch or a hearing aid. At the other end of the size scale, nobody would want a rechargeable battery for a guided missile, as it will be used once and destroyed in the explosion. Similarly, there are applications where a secondary battery is invariably used, *e.g.* to start cars, to power portable electronic devices (such as mobile telephones, laptop computers and camcorders), to provide stand-by (emergency) electricity supplies, and to propel electric vehicles.

In the home, however, there are situations where the choice between primary and secondary batteries is not so clear-cut and either may be used. To allow inter-changeability, small consumer cells come in standard sizes, *e.g.* AAA, AA, C, D, PP3, irrespective of the chemistry (see Figure 5.1, Chapter 5). Unfortunately, this inter-changeability does not extend to voltage. Most primary cells, *viz.* zinc–carbon or alkaline-manganese, generate 1.5 V (see Sections 5.1 and 5.2, Chapter 5), whereas secondary cells commonly used in the same applications, *viz.* nickel–cadmium and nickel–metal-hydride, generate 1.2 V (see Sections 9.4 and 9.6, Chapter 9). It follows that a 6 V battery pack requires five secondary cells, but only four primary cells. If primaries are replaced by secondaries, the terminal voltage will fall from 6 V to 4.8 V and there will be a corresponding limitation in performance. Modern lithium batteries confuse the situation further as they generate ~ 3.5 V. One way to overcome the problem of differing voltage and to take advantage of rechargeability is to employ rechargeable alkaline-manganese (RAM) cells (see Section 9.1, Chapter 9). These are cheaper than conventional nickel-based secondary cells and can provide a useful, but restricted, number of charge–discharge cycles.

Aside from the voltage consideration, the other factors which have to be taken into account are as follows.

- *Cost*: the outlay for rechargeable batteries is greater than that for primaries, and there is also the charger to be bought. Nevertheless, these extra costs are offset after just a few recharges, and thereafter the use of rechargeables is sheer profit. Frequency of use is therefore a key factor in deciding whether to purchase a primary or a secondary battery.
- *Convenience*: some people find it bothersome to have to remember to recharge batteries. Also, whereas the voltage decline of a primary cell is gradual and there is adequate warning that it is reaching time for replacement (*e.g.* the dimming of a torch light), secondary cells have an abrupt end-point of discharge and give very little notice of the need to recharge. Both inconveniences may be met, at some additional cost, by purchasing a spare set of batteries, so that one set is being recharged while the other is in use.
- *Capacity*: alkaline-manganese primary cells have a capacity which is

four times higher than that of nickel–cadmium rechargeable cells, as well as a higher voltage (see Table 9.1, Chapter 9).

The relative importance of these factors will depend upon the battery application.

3.2 ENERGY OR POWER?

When comparing the energy content of batteries of broadly similar voltage, such as 1.5 V primary batteries or 1.2 V nickel secondary batteries, it is common practice simply to state the capacity of the battery expressed in ampere-hours or milliampere-hours. Most manufacturers' data sheets use this convention; the rated capacity is quoted in terms of discharge at a given rate and temperature to a specified cut-off voltage. It is important to note these qualifications, as the capacity actually delivered will depend upon all three conditions. Table 3.1 shows the rate-dependence of the capacity of an alkaline-manganese cell (D-size) discharged at 21 °C to a voltage of 0.8 V. In the case of primary batteries, as mentioned in Section 2.2, Chapter 2, it is more usual to quote the discharge rate in terms of the available service hours of discharge through a load of stated resistance, rather than as a constant current which can be sustained for a given time. This practice is mostly a matter of user convenience.

To compare the energy content of batteries of differing voltage, it is necessary to multiply the capacity by the voltage to obtain the energy content expressed in watt-hours (Wh). For this purpose, it is conventional to use the open-circuit voltage as this gives the intrinsic energy stored in the battery. The energy delivered to the load will depend, however, on the discharge rate since, as we have seen, the on-line voltage, as well as the available capacity, falls with increasing discharge rate.

The energy content of a battery is not too meaningful, except in the context of the mass and the volume of the battery. In general, both of these parameters are important to the design engineer and the battery

Table 3.1 *Capacity delivered by an alkaline-manganese cell at various discharge rates (D-size cell; temperature: 21 °C; cut-off voltage: 0.8 V)*

Current (mA)	Capacity (Ah)
115	14.4
235	11.7
475	9

user, although their relative importance varies with the application. For instance, one of the major disadvantages of the lead–acid secondary battery is its mass, although this is much more significant for portable applications than for stationary operations, such as the supply of emergency power from stand-by batteries. In other uses, the volume of the battery assumes an importance which is at least equal to that of its mass. Comparison between batteries in these respects is made by defining the terms:

- **specific energy** (or **gravimetric energy density**), which is the stored energy per unit mass, expressed as watt-hours per kilogram of mass $(Wh\ kg^{-1})$;
- **energy density** (or **volumetric energy density**), which is the stored energy per unit volume, expressed as watt-hours per litre of volume $(Wh\ l^{-1}$ or $Wh\ dm^{-3})$.

For small consumer batteries, the manufacturers often use capacity rather than energy to qualify their products, and specify the corresponding capacity densities in smaller units, *e.g.* $mAh\ g^{-1}$ or $mAh\ cm^{-3}$.

Another term which is commonly employed is the **theoretical specific energy** of a battery. This is calculated from the reversible cell voltage, the number of electrons transferred in the cell reaction and the masses of the reactive chemicals (the active materials), and takes no account of the mass of all the other inert components in the battery, *e.g.* cell container, terminals, current-collectors, busbars, cell inter-connectors, separators, electrolyte. Note, the calculation for the lead–acid battery is unique in that it includes the electrolyte (sulfuric acid) which takes part in the cell reaction [see Equation (2.5), Chapter 2]. The **practical specific energy** of a battery is always much lower than the theoretical value, often by a factor of four or five. This is accounted for partly by the construction materials, which add to the battery mass but are not involved in the energy-producing reaction, and partly by electrokinetic and other restrictions which serve to reduce the cell voltage and prevent full utilization of the reactants. The usefulness of theoretical specific energy is to allow a comparison of the prospects for different cell chemistries before ever building any batteries, although with the understanding that this concept is an ideal which will not be approached in practice.

A second major concern of many battery users is the **power** available from the battery, expressed in watts (joules per second). Similar to the above terms for energy, **specific power** and **power density** relate to the power output per unit mass $(W\ kg^{-1})$ and the power output per unit volume $(W\ l^{-1}$ or $W\ dm^{-3})$, respectively. Depending upon the application, the requirement may be for continuous power, for peak power over a short period (say, 1 to 10 min), or for power pulses of a second or a sub-second duration. The power output of a battery is determined as much by its design and construction as by its chemistry; certainly, for a given cell chemistry, it is possible to design cells of widely differing power output.

Unfortunately, there is usually a trade-off between power output and stored energy, namely: the higher the power available, the lower the stored energy. The fundamental reason for this trade-off is that high power output requires a cell of low internal resistance and low electrode polarization. This, in turn, calls for thin electrodes of high surface area, packed closely together. The result is that the inert current-collectors, separators, *etc.* constitute a higher fraction of the mass and volume of the cell, while the active materials represent a smaller fraction. Thus, the specific energy and energy density of a high-power cell are less than those of a conventional cell of the same chemistry. By way of example, the specific energies and specific peak-power ratings (at 80% DoD) of two nickel–cadmium traction batteries are compared in Table 3.2. One battery is in the standard form, while the other is an extra-high-power version. The latter has 83% more peak power, at the expense of 36% less stored energy. In this particular example, the high-power battery is intended for use in a hybrid electric vehicle, that is one in which a small internal combustion engine provides range and the battery supplies a power boost for short periods when accelerating, overtaking, or hill-climbing. The standard version of the battery is more appropriate to a conventional electric vehicle, where it has to provide range as well as power.

Design engineers are often as much constrained by the available space in which to fit a battery as they are by the permissible weight. It follows that all four parameters, *viz.* specific energy (Wh kg^{-1}), specific power (W kg^{-1}), energy density (Wh dm^{-3}) and power density (W dm^{-3}), are of interest to them. This is particularly true for electric-vehicle batteries. As

Table 3.2 *Specific energies and specific peak-power ratings of nickel–cadmium traction batteries*

	Standard cell	*High-power cell*
Rated capacity (Ah)	100	56
Specific energy (Wh kg^{-1})	55	35
Specific peak-power (W kg^{-1})[a]	120	220

[a] Measured at 80% DoD.

we shall see in Chapter 8, the lead–acid battery is widely used as a traction battery as its volume is such that quite respectable energy and power densities can be attained; it is only in regard to mass that it is seriously deficient.

Another sphere in which battery power assumes key importance is in military applications such as guided weapons and missiles, where mass and volume are prime considerations and the mission is of short duration. The requirement, therefore, is for a lightweight and compact battery of high power output.

It may readily be shown, by simple physics, that the peak-power output from a battery is obtained when the load voltage is equal to half the open-circuit voltage, *i.e.* when the resistance of the load is equal to the internal resistance of the battery. This assumes that the internal resistance of the battery is a constant, which is not strictly correct because of increasing polarization and thermal effects that arise with increasing temperature. Nevertheless, it is a valid first approximation. In practice, the maximum *continuous power* output from a battery is often limited by thermal considerations. This is especially true for large monoblocs and for battery packs in which the monoblocs are packed closely together. Electrical inefficiencies (see below) lead to the generation of heat, and the rate at which this heat can be dissipated is the limiting factor in the continuous power rating of the battery. Some battery packs are cooled by convection or forced-air circulation, or even by circulating liquid coolant. For *pulsed power* discharge, thermal effects are by no means as severe and much higher discharge rates are possible for short periods. Some of the newer lithium-ion cells have exceptionally high pulse-power output. A small 6.5 Ah cell is capable of giving power pulses of $1.5\,\text{kW}\,\text{kg}^{-1}$ ($3\,\text{kW}\,\text{dm}^{-3}$) when fully charged (see Table 10.1, Chapter 10). Clearly, such power densities could not be sustained for long without damaging the cell.

3.3 BATTERY TEMPERATURE

A further important consideration in choosing a battery is the range of ambient temperature over which it is expected to perform satisfactorily. Here, we may distinguish between indoor use and outdoor use. The interiors of buildings are generally kept at 15 to 30 °C for reasons of personal comfort. This is also a temperature range which is comfortable for most batteries. By contrast, outdoor temperatures may extend from $-40\,°\text{C}$ or lower in continental winters to $+70\,°\text{C}$ or higher in an enclosed container (such as a vehicle) in tropical summers. Such a range is comfortable neither for humans nor for most types of battery, and electrical performance generally falls off sharply at these extremes of temperature, particularly at the lower temperatures. To demonstrate this feature, the specific energies of three primary cells, all with zinc negative electrodes, are compared in Figure 3.1. In each case, the maximum energy is obtained at about 40 °C. At temperatures below 0 °C, the zinc–mercury-oxide and zinc–carbon cells become almost useless, while the performance of the alkaline-manganese cell is also declining, though less sharply. Other primary cells, such as lithium–manganese-dioxide with an organic electrolyte, are useable down to − 40 °C, although again with a sharp fall-off in performance (Figure 3.2).

Rechargeable batteries also may have to operate at low temperatures. A particular situation where high power output is required at low tem-

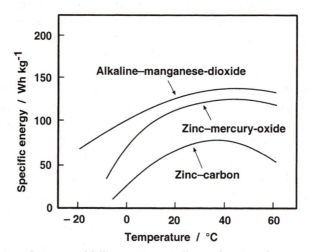

Figure 3.1 *Specific energy of different primary cells as a function of temperature* (By courtesy of Duracell Batteries)

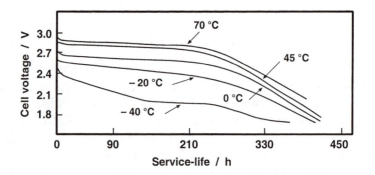

Figure 3.2 *Discharge curves for a lithium–manganese-dioxide cell* (By courtesy of Ultralife Batteries, Inc.)

perature is in the starting of internal combustion engines in cold climates. Under such conditions, not only does the performance of the usual lead–acid automotive battery decline, but also more current is required to turn over the engine because of the increased viscosity of the oil. Obviously, battery systems which function only when hot are unaffected by extremes of ambient temperature. Rechargeable sodium batteries (see Sections 11.5 and 11.6, Chapter 11), for example, work at 200 to 400°C and it is necessary, in fact, to contain them in a heated, thermal enclosure.

3.4 SHELF-LIFE

Most batteries, particularly those with an aqueous electrolyte, undergo slow self-discharge when left standing on open-circuit. The rate at which this occurs depends on the cell chemistry, its construction and, especially, on the temperature of storage. Usually, the rate of self-discharge increases with temperature.

Shelf-life is more of a concern for primary batteries than for secondary batteries, as the latter can always be recharged before use. Modern alkaline-manganese primary cells have a storage life of several years at 20°C, which is adequate for most purposes, although storage at high ambient temperatures could create a problem. Lithium–manganese-dioxide batteries have an even longer shelf-life (see Sections 6.2 and 6.3, Chapter 6). One area where particularly long shelf-lives are sought is in the defence field as weapons may be in storage for many years before being called upon to operate reliably at a moment's notice.

3.5 ENERGY EFFICIENCY AND RECHARGE RATE

On discharging a secondary battery, the amount of electricity recovered, measured in watt-hours, is always less than that used in charging the battery. The ratio watt-hours recovered : watt-hours introduced is the **electrical** (or **energy**) **efficiency** of the charge–discharge cycle. The electrical loss associated with the cycle is made up of two components:

- **Coulombic inefficiency**. This arises from electric current wasted in non-productive side-reactions, such as decomposition of the electrolyte solution ('gassing'), corrosion of battery components, *etc.*
- **Voltaic inefficiency**. As shown in Figure 2.3, the voltage required to charge a secondary battery is always greater than the discharge voltage, and the difference, $V_{ch} - V_d$, is a measure of the voltaic inefficiency of the overall process. This inefficiency results from the internal resistance of the cell and from polarization losses at the

electrodes. The greater the current flowing during charge and discharge, the more the on-line voltage departs from the reversible voltage of the cell, V_r, and the greater is the voltaic inefficiency (see, for example, the charge–discharge behaviour of a sodium–metal-chloride cell, Figure 2.4, Chapter 2).

The product of the coulombic efficiency and the voltaic efficiency gives the overall energy efficiency of the battery under the given conditions of operation. The resulting value varies not only with the rates of charge and discharge and with temperature, but also within any given charge or discharge half-cycle because the losses do not remain constant. For instance, as the charging reaction approaches top-of-charge, more gas is released and the coulombic efficiency falls. At the same time, polarization increases, leading to a reduction in voltaic efficiency. Finally, as the battery moves into the overcharge regime, when only decomposition of the electrolyte solution is taking place, the coulombic efficiency falls to zero. Similarly, though less dramatically, overdischarge of a battery can lead to side-reactions and a decline in overall energy efficiency.

A sweeping generalization is that, under restrained conditions of operation, the coulombic efficiency of a secondary battery is likely to be $\sim 90\%$ and the overall energy efficiency 50 to 75%, as dictated by the charge and discharge rates employed. For many users, considerations of energy efficiency are unimportant as electricity is sufficiently cheap and other factors weigh far more heavily in choosing a secondary battery. There are situations, however, where overall energy efficiency does become important, as shown by the following examples.

- When the battery is located at a remote site, or one that is difficult to access. Electrolyte lost through gassing has to be replaced periodically and this maintenance operation needs to be as infrequent as possible. In this situation, a sealed, 'maintenance-free' battery is often employed.
- When the batteries are large and the daily consumption of electricity is substantial as, for example, when running a fleet of commercial electric vehicles. In such operations, the cost of electricity may represent a significant component of the overall operating cost.
- When the battery provides the energy-storage component for a solar array or other renewable energy source. Solar arrays are still very expensive, although reducing in price, and economic considerations dictate that the battery should be as efficient as possible to avoid wastage of costly solar electricity.

Another important consideration is the time taken to recharge a secondary battery. The charge rates recommended by most battery manufacturers are substantially less than the discharge rates which are commonly employed. Frequently, a recharge rate of 8 to 10 h is recommended and, in practice, many batteries are charged overnight. There are circumstances, however, when a fast recharge is desirable. For instance, the owner may have forgotten to switch on the charger overnight, a fuse may have blown, or the mains may have failed. Another potential use for fast charging is to extend the effective range of electric vehicles. In the USA, public charging facilities for electric vehicles are now being installed in California and other States, while recent research has shown that it is possible to charge lead–acid batteries without damage at a much faster rate than previously thought possible (see Section 4.4, Chapter 4). Of course, such fast charging is not compatible with high energy efficiency, although the worst of the inefficiency can be mitigated if the battery is not fast charged to more than 80% of its nominal capacity.

3.6   BATTERY LIFE

The life of a primary battery is normally defined by an end-point voltage, on the steep part of the discharge curve. Once discharged beyond this point, the cell is discarded.

Secondary batteries may either fail precipitously and catastrophically, or exhibit a gradual decline in performance until no longer useful. In the latter case, the operational life of a secondary battery may be defined chronologically (years of use), or in terms of the number of charge–discharge cycles that it will sustain, termed the **battery cycle-life**. The **end-of-life** may be taken as the point at which the capacity of the battery has fallen to an unacceptable level, or at which its maximum power capability at 80% DoD has fallen to 80% of its initial value. These are practical criteria for which battery performance is still acceptable until the end-of-life. Depending upon the application and the discharge regime being used, either of the criteria may be considered to be more important than the other.

The cycle-life of a secondary battery is critically dependent upon the cycling regime to which it has been exposed, *i.e.* its **cycling history**. Deep-discharge cycles and overcharging are especially detrimental and shorten the cycle-life considerably. By contrast, a battery is usually able to withstand a much greater number of shallow discharge cycles. Fortunately, some of the uses for secondary batteries involve only shallow cycles (*e.g.* engine starting), or only very occasional deep discharges (*e.g.* stand-by power supplies). Often, the battery is little more than an energy-storage buffer between the charger and the load. For example, automotive batteries are constantly recharged during driving and the battery is a buffer between the alternator and the various electrical loads in the car.

In other situations, where regular deep discharge might be expected, shallow cycling is often preferred for operational reasons. For example, traction batteries for industrial tractors and fork-lift trucks may be partially recharged during idle periods throughout the day so as to allow them to work longer shifts. When regular deep discharge really is necessary, most manufacturers recommend that it does not exceed 80% of the nominal capacity, in order not to shorten battery life excessively. Allowance must be made for this procedure when estimating the size of battery required for a particular duty cycle. Indeed, there is likely to be a certain trade-off between the capital cost and the life of the battery. Excess capacity (*i.e.* more that required for the duty cycle), results in shallower cycling and longer life for the battery, and also provides a reserve for those occasions when the duty cycle is greater than normal. These advantages must be carefully weighed against the disadvantages of added cost, mass and volume.

In summary, the choice of a battery for a particular application is by no means straightforward and involves many considerations of which the most important are voltage, capacity (size), desired technical performance, and cost. The technical performance sought depends upon the application for which the battery is to be used and the anticipated duty cycle. The specification may be divided into vital attributes, which are indispensable, and desirable attributes which may have to be traded, in some measure, against capacity and cost. By way of illustration, we conclude this discussion by listing the principal features which may be required in a secondary storage battery.

- A stable voltage plateau over a good depth-of-discharge.
- High specific energy (Wh kg^{-1}) and high energy density (Wh dm^{-3}).
- High peak-power output per unit mass (W kg^{-1}) and volume (W dm^{-3}).
- Wide temperature range of operation.
- High energy efficiency (Wh output : Wh input).
- Long cycle-life with deep-discharge cycling.
- Ability to accept fast recharge.
- Good charge retention on open-circuit stand.
- Ability to withstand overcharge and overdischarge.
- Reliable in operation.
- Maintenance-free.
- Rugged and resistant to abuse.

- Safe both in use and under accident conditions.
- Made of readily available materials which are environmentally benign.
- Suitable for recycling.

Overall, this is a formidable general specification, even before going into technical detail, and explains why the development of new batteries has proved to be so difficult. One of the complications is that many of these parameters are interrelated in a complex fashion. For example, recoverable energy depends on the discharge rate and the ambient temperature, peak-power output on the state-of-charge of the battery, and cycle-life on the depth-of-discharge. Batteries with different chemistries vary widely in their performance parameters, but so also do different designs of battery with the same chemistry. It is this latter fact which makes it so difficult to make precise comparisons of the various parameters for different battery systems. Although such comparisons have been widely made, they are at best only semi-quantitative. This is because many of the numerical values depend as much on the battery design and on the conditions under which the battery is tested as on its fundamental chemistry.

Chapter 4

How to Charge a Secondary Battery

The charging of a secondary battery involves passing direct current (d.c.) electricity through the battery in order to reverse the discharge process. An electrochemical reduction reaction ('electroreduction') takes place at the negative electrode and generally leads to the formation of a metal, while electrochemical oxidation ('electrooxidation') of the active material occurs at the positive electrode. The charging of a battery is therefore an electrolytic process. The conditions under which a secondary battery is charged are at least as important as the discharge conditions in determining its useful life, and yet charging is a much neglected half of the electrochemical cycle. This chapter sets out to redress this neglect. The emphasis is on charging techniques for lead–acid batteries since these are by far the most common type of rechargeable battery, especially in the larger sizes.

The 'charge-acceptance' (C_A) of a battery is defined as that part of the current flowing through the battery that is utilized for the charging process. It is quantified as the ratio, expressed as a percentage, of the number of ampere-hours that are usefully accepted during a small increment of time to the total number of ampere-hours supplied in that time. Thus:

$$C_A = 100(I - I_g)/I \tag{4.1}$$

where I is the total current flowing and I_g is the current wasted in gassing or other side-reactions.

If current continues to flow after recharge is complete, then the battery enters the overcharge regime. 'Overcharge' of a battery is said to occur when the active material of one of the electrode polarities (the material which is present in deficit quantity) is fully converted into the charged state. Batteries may be built with either an excess of negative material

34

('positive-limited') or an excess of positive material ('negative-limited'). On charging the positive-limited design, the positive active-material will obviously be the first to approach full conversion. Correspondingly, the positive electrode will be taken to a greater depth-of-discharge than the negative during battery operation. Such duty can result in degradation of the positive electrode. This is particularly true for lead–acid batteries because the molar volume of the discharge product (lead sulfate) is $\sim 70\%$ greater than that of the active material (lead dioxide). The resulting mechanical stresses cause weakening and disintegration of the paste structure of the positive plate (so-called 'paste softening'), an expansion process which progresses through to shedding of active material and, therefore, to loss of plate capacity. (Note, lead crystals formed in the negative plate during charging tend to coalesce and thus promote shrinkage, rather than expansion, of the active material, see Section 8.3.3, Chapter 8.) Accordingly, there is a preference within the industry for the negative-limited design of battery in which the positive plates are subjected to a reduced depth-of-discharge and are therefore 'protected' from material instability problems during the course of charge–discharge cycling.

Overcharging will result in the electrolyte being decomposed electrochemically. For batteries based on aqueous electrolytes, decomposition leads to the evolution of oxygen and hydrogen. Such gassing is undesirable for a number of reasons: (i) it wastes electricity; (ii) the water lost by electrolysis has to be replaced, which necessitates more frequent maintenance; (iii) it presents a certain safety hazard; (iv) the formation of gas bubbles weakens the structure of the active material, particularly at high current densities, and thereby reduces the cycle-life of the battery.

The low-voltage, direct current required to charge a battery is usually supplied from one of two sources. For stationary batteries and for electric-vehicle traction batteries, mains electricity is used. This is fed through an installed charger which is a combined transformer and rectifier. On the other hand, if the battery is operating in a conventional road vehicle, train, boat or aeroplane, the source of electricity is an alternator driven by the internal combustion engine. This provides a low-voltage, alternating current (a.c.) output which is rectified, generally by a diode circuit which is integral with the alternator. Sometimes, when mains electricity is not available, stationary batteries are charged by a small, portable, internal combustion engine and d.c. generator set. These sets are often seen on building sites and at road works where, apart from charging batteries, they provide a local electricity supply for powering tools, for running temporary traffic lights, *etc.* Moored boats frequently have auxiliary d.c. generator sets to charge the crafts' batteries while stationary. A relatively new source of power, which is likely to assume greater significance in future, is solar electricity. Installed solar panels feed storage batteries by day, so as to make available a source of electricity after dark. As the electrical output of a solar array varies with the season, the time of day and the extent of cloud cover, it is particularly critical to provide good control of the battery charging voltage. Batteries may also be charged by wind generators and this is often the preferred route for small boats and caravans.

As has been noted, serious overcharge of a battery is highly undesirable because it shortens cycle-life and also may lead to a requirement for increased maintenance through loss of electrolyte *via* gassing. The key to avoiding overcharge is to have accurate control of the charging voltage, especially the top-of-charge voltage. Most charging circuits include a voltage regulator which is adjusted to the highest voltage appropriate to the battery being charged. So-called 'temperature compensation' may also be required. For example, the top-of-charge voltage of lead–acid batteries decreases with increasing temperature and hence the voltage-regulator should be programmed to provide a given rate of voltage change with temperature.

4.1   CHARGING REGIMES

Four basic regimes are used for the charging of secondary batteries, namely: constant-current charging, constant-voltage charging, taper-current charging, and constant-current–constant-voltage charging. These regimes are described in the following sections.

4.1.1   Constant-current Charging

As the name implies, this regime involves using a constant current throughout the charging process. The procedure can be achieved with chargers which are both reliable and inexpensive, but there is a problem in deciding the level of current to use. If the current is too low, then the overall charging process is very slow [Figure 4.1(a)]; if a higher current is used, then the gassing at top-of-charge becomes excessive. Ideally, a high current is employed in the first half (or more) of the charging process, followed by a smaller current in the later stages. This leads to the concept of the two-step, constant-current charge [Figure 4.1(b)] in which a medium-to-high rate charge is used in the first period, followed by a very low 'trickle' rate as the battery approaches full charge. The advantage of using a small trickle-charge over an extended period is that it allows batteries in a long series-connected chain to be rebalanced at top-of-charge without

(a) Constant-current: single-step

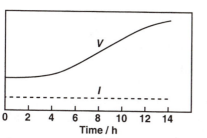

(b) Constant-current: two-step

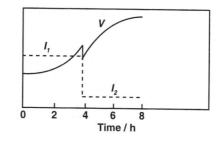

(c) Constant-voltage

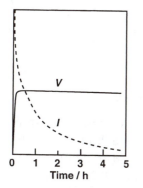

(d) Taper current: single-step

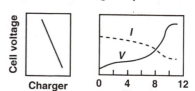

(e) Constant-current – constant-voltage

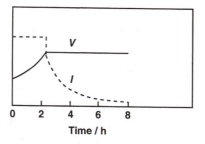

Figure 4.1   *Various procedures for charging batteries*
(By courtesy of Research Studies Press Ltd)

causing serious damage to those batteries (or cells) which are already fully charged. Cell balancing improves both the performance and the life of the battery. Constant-current charging, whether single-step or two-step, is a relatively slow process and is best suited to overnight charging.

4.1.2   Constant-voltage Charging

When charging at constant-voltage, the level of current supplied is determined by the voltage difference between the charger and the battery. The
current starts off at a very high value and then decreases roughly exponentially as charging proceeds [Figure 4.1(c)]. As a consequence, it
takes a long time to achieve a full charge. A further disadvantage is that if
the battery is deeply discharged, the starting current can be very high
indeed, which necessitates the use of a large, expensive charger and may
give rise to unacceptable internal heating of the battery. For these reasons, constant-voltage charging is usually confined to batteries which
experience relatively shallow discharges, such as those employed in stationary 'float' duties, where long-term (continuous) trickle-charging
would result in excessive loss of electrolyte by gassing. Constant-voltage
chargers may also be used as fast chargers to return maximum charge in
minimum time.

4.1.3   Taper-current Charging

A taper charger is an inexpensive design in which the current starts off
high and decreases progressively as the cell voltage increases [Figure
4.1(d)]. This is an unregulated charger as the end of charge is usually
controlled by a fixed voltage rather than by a fixed current. Many cheap,
domestic chargers for use with car starter batteries are of this type, with
the maximum current limited to a few amperes.

4.1.4   Constant-current–Constant-voltage Charging

In this charging regime, the current is held constant until the battery
voltage reaches a pre-defined value where gassing is likely to begin
[Figure 4.1(e)]. At this point, the voltage is held constant and the current
allowed to decline exponentially, as with constant-voltage charging.

   In a battery pack of series-connected cells, problems may arise as
charge–discharge cycling progresses. These problems stem from inherent
differences and imbalances between the cells in a string. When the battery
pack is new there may be slight differences in the internal resistance
and/or the capacity of the individual cells. These differences are due to
slight variations in cell construction and assembly. Moreover, as the cells
age, their capacity fades and their resistance increases. This process is
unlikely to be exactly uniform from cell to cell so that some cells in the
chain will become weaker than others. A weak cell will be more heavily
drained than stronger cells and, on charging, will require more time to

reach full charge. If the charge is inadequate and the battery is discharged again, then the low-capacity cell will be even more heavily discharged. Should this process be allowed to continue indefinitely, the weak cell will become less and less charged and, ultimately, will become 'dead', or even go into reversal (where the polarity of the electrodes switches). Such 'cycling-down' behaviour can be controlled by subjecting the cell string to periodic 'equalizing' (or 'levelling') charges by maintaining the top-of-charge voltage for twice the usual time until the weaker cells are fully charged. This procedure needs to be repeated every ten cycles or so, but if any cell is seriously defective it will ultimately need to be replaced.

There are also charging idiosyncrasies which depend upon the chemistry of the cell. For example, it is well known that some nickel–cadmium cells (see Section 9.4, Chapter 9) exhibit the so-called 'memory effect' whereby if they are consistently subjected to only light discharges, followed by recharge, after a while they seem incapable of deep discharge. The 'lost' capacity may sometimes be restored by first shorting out the terminals of the nominally discharged cell, so as to ensure that a fully-discharged state is reached, and then recharging the cell. Another example is the sodium–sulfur cell (see Section 11.5, Chapter 11). This cell has no overcharge mechanism, so that when the top-of-charge is reached, the cell becomes infinitely resistive. If the cell is a weak member of a series-chain of such cells, then the total voltage of the chain will fall across this one cell. This will cause dielectric breakdown of the solid ceramic electrolyte and, thus, catastrophic failure of the cell. The problem may be avoided by constructing the battery from a series–parallel matrix of cells with parallel connections every few cells, so that the fully-charged weak cell is effectively by-passed. From these examples, it is clear that it is necessary to be aware of any peculiarity which arises from the chemistry of the battery being charged.

4.2 TYPES OF CHARGER (DOMESTIC, INDUSTRIAL)

Almost everyone is familiar with the small, domestic charger supplied for use with consumer rechargeable cells and plugged into the mains electricity supply (Figure 4.2). Typically, this is used to charge nickel–cadmium or nickel–metal-hydride cells removed from portable compact-disc players, children's toys, *etc*. (It is also possible to recharge certain types of alkaline–manganese 'primary' cell, see Section 9.1, Chapter 9.) With other portable equipment (mobile telephones, workshop tools, *etc*.) the battery is not normally removed from the device which is fitted with a jack socket to accommodate the supply from the dedicated charger. Still other consumer appliances, for example electric toothbrushes powered by

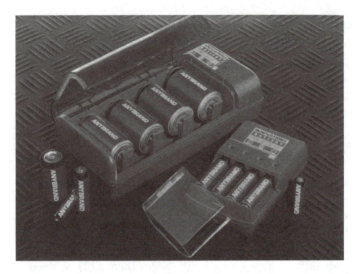

Figure 4.2   *Domestic chargers for consumer cells*
(By courtesy of Batteries International)

nickel–cadmium batteries, may be charged by 'inductive charging' (see
Section 4.4). Lithium-ion consumer batteries are particularly sensitive to
voltage control during charging and a special charger is needed (see
Section 10.4, Chapter 10).

   Chargers for larger secondary batteries, such as 12 V, 40 to 90 Ah
automotive or 'leisure' batteries, are readily available from car accessory
outlets at modest prices. The cheapest of these are taper-type chargers
which give a maximum of 5 A, but often only 1 to 2 A at top-of-charge. It
is also possible to purchase heavy-duty versions which provide up to 50 A
for fast charging, although it is not desirable to pass as high a current as
this in the overcharge regime.

   In industry, much larger units are employed to charge the traction
batteries used in such applications as fork-lift trucks, tugs and tractors, as
well as stationary batteries for uninterruptible power supplies. Often
these are quite simple charger units which conform to one of the above
charging regimes, although there is a growing tendency to replace older
units by computer-controlled chargers which monitor the state-of-charge
of the battery and determine the current to be employed. In the lead–acid
battery industry, the last stage of manufacture is to charge the batteries –
the so-called 'forming reaction'. Because the process takes some hours,
many batteries have to be formed together and this requires a considerable amount of electrical energy. Thus, chargers are sold which operate
up to several hundred volts and/or at high currents. An example of such a

charger unit, with which various standard charge profiles are possible, is given in Figure 4.3.

Some of the largest industrial chargers are those which have been developed for the fast charging of electric-vehicle traction batteries so as to enhance the daily range of the vehicle. A 300 kW rapid charger is shown in Figure 4.4. Clearly, this is a substantial (and costly) piece of engineering.

Lead–acid batteries (12 V) used for engine starting are on 'float charge' and are subjected only to shallow discharges. They are normally charged by an alternator with an output voltage held at 14.2 ± 0.4 V. It may be necessary, however, to adjust the voltage limit according to the type of lead–acid battery employed (see Section 8.3.5, Chapter 8). In addition, it has become important to provide temperature compensation given the trend towards hotter engine-compartments. This is because automobiles are being aerodynamically styled at the expense of underbonnet space which, in turn, is being increasingly packed with mechanical, electrical and electronic devices. Both these factors preclude air movement and thus inhibit the transfer of heat out of the engine compartment.

Charging is more complex when batteries are subject to regular deep discharge, as with the 'leisure' batteries used for lighting, water pumps, *etc.*, in caravans and boats. Commercial vehicles also may require deep

Figure 4.3 *Modern charger for the formation of lead–acid battery plates*
(By courtesy of Bitrode Limited.)

Figure 4.4   *300 kW rapid charger for charging electric buses*
                (By courtesy of Norvic, Inc.)

discharge, *e.g.* fire engines, trucks with electric tail-lifts, ambulances. Under such service conditions, where much larger currents flow during charge, the voltage losses within the battery and the charging cables, and across the diode, together with a lack of temperature compensation, may result in the battery being only 60 to 70% charged at the cut-off voltage. Electronic voltage regulators have been developed which compensate for these effects by continuously monitoring the relationship between the battery input and the alternator output voltages. The regulator thus ensures that the battery really does attain its correct set-voltage point and, therefore, its full charge. Numerous advantages are claimed for the use of such devices in terms of both battery performance and battery life, although these advantages have to be offset against the added cost of electronic voltage regulation.

A fairly recent development in charger design is the 'pulse charger' or 'switched-mode charger'. This came about as a result of better under-standing of battery chemistry by electronics engineers. When a battery is undergoing charge, there is a build-up of concentration gradients of ions in the vicinity of the electrodes. This results in electrode polarization and a drop in current for a given applied voltage. The concentration gradients may be dispersed by a short reverse (discharge) pulse, after which the current reverts to a higher value. Depending on the state-of-charge of the battery, discharge pulses can also serve to minimize gassing and to assist the dissipation of heat. It has been found to be further beneficial to use a series of short, high amplitude, current pulses during charge, rather than a lower, continuous current. This activates more nucleation sites for reaction to take place on the electrode surface, which has the favourable and simultaneous effects of substantially increasing the charge current and reducing the likelihood of the formation of metallic dendrites. The latter are metal crystals, in the form of 'spikes', which grow out from the negative electrode surface and can penetrate the separator to give rise to internal short-circuits. Diffusion during the rest periods between the current pulses assists the dissipation of concentration gradients. The charging regime therefore consists of a series of high-current charge pulses, with rest periods between, and a periodic reverse (discharge) pulse to remove excess build-up of ions in the vicinity of the electrodes. The application of pulse-charging techniques is still in its early days and is largely confined to small/medium-sized batteries, especially rechargeable nickel-based batteries. This is because the complex power electronics required by the chargers makes large units somewhat expensive.

4.3   CHARGING INDUSTRIAL TRACTION BATTERIES: OPPORTUNITY CHARGING

By far the majority of electric vehicles are off-road vehicles. Electric tugs and tractors are widely used in airports, railway stations, mines, city parks, and in any situation where the pollution and noise associated with internal combustion engines are unacceptable. Electric golf buggies are similarly chosen, in part, for environmental reasons. In industry, vehicles for materials' handling (fork-lift trucks, pallet trucks, *etc.*) are generally electrically propelled when they have to be used inside buildings.

Traditionally, the traction batteries for such vehicles have been charged overnight (*i.e.* once in 24 h), but there is a growing realization of the advantages of using 'opportunity charging'. This simply involves plugging into the charger whenever there is an opportunity, *e.g.* between shifts, when waiting for work, during rest or meal breaks. Such charging is claimed to provide the following benefits:

- savings on capital cost – no second battery is needed for working two shifts or overtime;
- savings in labour costs – there is no requirement to change batteries between shifts;
- a smaller and cheaper battery will suffice for single-shift operation.

On the deficit side, daytime electricity tariffs are often higher than off-peak tariffs, and it may be necessary to have multiple charging sites scattered around the factory as well as one central facility for overnight charging.

Opportunity charging works best when batteries are in a 30 to 70% state-of-charge. It is important to avoid overcharging and excessive gassing, by using an appropriate charging regime, and not to allow discharge below the usual limit of 80% DoD. As a practical example of opportunity charging in action, Figure 4.5 shows data for an airport baggage tractor operated intermittently from 6 a.m. to 11 p.m., followed by an overnight charge for 7 h. Without opportunity charging, it is necessary to change the battery in the early afternoon. With opportunity charging, a single battery will suffice, even though 140% of the battery capacity is used each day. An extended charge to balance the cells (an 'equalizing' charge) is carried out once a week.

At one factory with 120 electric trucks, it was reported that switching from overnight charging to opportunity charging resulted in a reduction of the number of batteries required from 290 to 135. Moreover, although the energy passing through each battery more than doubled, there was no observed decrease in average battery life. This success was achieved through the use of a computer-controlled battery management system which ensured that the batteries were not overcharged, or discharged too deeply, and also warned when overheating was taking place.

4.4   CHARGING ELECTRIC ROAD VEHICLE BATTERIES

Electric cars are now being introduced gradually into Europe, Japan, and the USA (particularly California). Most of the major car manufacturers in these regions have prototype electric vehicles (EVs), either vans, light trucks or small cars, and a number of them are setting up their first, small-scale production lines (see Section 12.4, Chapter 12). While it is

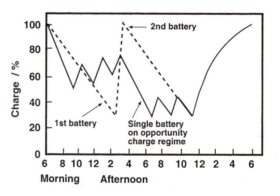

Figure 4.5   *Performance of an airport baggage tractor, with and without opportunity charging*
(By courtesy of Batteries International)

expected that EVs will charge their batteries overnight at their home base, the energy-storage limitations of existing batteries dictate the need for opportunity charging so as to enhance the daily driving range. This, in turn, leads to a requirement for public fast-charging stations dispersed over a wide area and situated at convenient locations such as workplaces, railway stations, shopping centres, *etc*. Realizing that the future sale of electric cars depends upon the availability of public charging stations, promoters are now installing significant numbers of such stations in Arizona and California in the USA. There is also some similar activity in Europe, particularly in France.

Recent research, backed by practical demonstrations, has shown that lead–acid traction batteries can be fast charged without serious detriment. The goal is to return 50% of the depleted charge in 5 min and 80% in 15 min. If these targets are attained systematically, then the concept of the 'refuelling site', analogous to the petrol station, becomes a practical reality. The car will be effectively recharged while the driver has a stop for refreshment in the associated coffee shop. Consider a family-sized car with a 25 kWh traction battery. If it is to be charged from 20 to 80% state-of-charge in 15 min, then a 60 kW d.c. supply is required, given that there are no electrical losses and a uniform rate of charging. If the charger is 80% efficient, the a.c. supply will need to be 75 kW. For an 'electrical service station' to be able to handle 20 vehicles at once (maximum of 80 per hour), a 1.5 MW supply must be available. This would be a substantial and costly undertaking. Nevertheless, the implementation of such facilities is receiving serious consideration.

There is still a major controversy over the type of charger to be employed with electric road vehicles. Most developers favour the tried and tested 'conductance charger' in which there is a physical electrical connection (a cable) between the a.c. supply, the charger, and the vehicle. The charger is likely to be an off-vehicle, stationary installation, although smaller vehicles may carry their own portable charger for low-rate charging. The advantage of the latter approach is that recharging may be carried out wherever there is a convenient a.c. electricity outlet socket. The disadvantage is that, as the supply is usually limited to 13 or 15 A and the charger is small, recharge will be restricted to a slow rate (say, 8 to 12 h for a full charge). Fast charging requires a large, ground-based charger. One problem with conductance charging is that there is, as yet, no internationally agreed standard for the design of the connector to carry high currents safely under all operational conditions.

The alternative type of fast charger is the 'inductance charger'. In this device, there is no physical connection between the supply and the battery. Rather, a coil through which a high-frequency current flows is brought into close proximity with another coil on board the vehicle and a current is induced in the latter to charge the battery. This concept has been brought to practical realization by the Hughes Corporation in the USA. Instead of a plug and socket, the primary coil is contained in a 'paddle', about the size and shape of a table-tennis bat, and this is inserted into a slot in the electric car (Figure 4.6). The off-board Magne-Charge® control cabinet (Figure 4.7) converts a.c. mains power into high-frequency a.c. (80 to 350 kHz), which passes through the paddle and induces a current in the secondary coil on board the vehicle. This is then transformed and rectified to low-voltage d.c. for battery charging. The induction paddle may also contain auxiliary high-frequency, low-current coils for transmitting data from the battery management system to the charger.

In California, conductive charging stations and inductive charging stations are available at about 200 and 300 public locations, respectively. It is claimed that the inductive charging system is easy to use in all weather conditions. The driver finds it simpler to insert a paddle into a slot than to make a more complex physical connection to carry hundreds of amperes, as required for fast charging. At the present time, the car manufacturers are polarized in their support of one system or the other: General Motors, Nissan and Toyota have favoured inductive charging, while Chrysler, Ford and Honda have preferred conductive charging. Obviously, this situation needs to be resolved and a common standard adopted world-wide.

Figure 4.6 *Inserting the inductive paddle into the charging slot of General Motors' EV1* (By courtesy of Batteries International)

Figure 4.7   *Magne-Charge® 50 kW inductive fast charger, as installed in California and Arizona*
(By courtesy of Batteries International)

4.5   BATTERY MONITORING AND MANAGEMENT SYSTEMS

Evaluating the state-of-charge of a battery is not always straightforward. When the profile of the discharge curve is rather steep, as with primary zinc cells (see Chapter 5) or secondary lithium-ion cells (see Section 10.4, Chapter 10), a simple measurement of the open-circuit voltage gives a reasonable indication of the state-of-charge. By contrast, when the discharge profile is somewhat flat until almost the end of discharge, as can be the case with nickel–cadmium cells (see Figure 2.6, Chapter 2), determination of the state-of-charge is more difficult. In the special case of lead–acid batteries, advantage is taken of the fact that the sulfuric acid electrolyte participates in the cell reaction and becomes more dilute as discharge proceeds. Estimation of the state-of-charge of a flooded lead–acid battery is therefore a relatively simple matter, namely, measurement of the relative density of the electrolyte by means of a hydrometer. Such hydrometers are readily available at car accessory stores. In the case of a 12 V monobloc, it is necessary to examine each of the six cells, to check that they are discharging uniformly. To obtain a reliable result, the battery should be left overnight before making the measurement to allow diffusion processes in the acid to eliminate concentration gradients which arise when the battery is discharged or charged. Although this is a simple procedure, it is, at best, inconvenient and labour-intensive, especially if there are many batteries to be checked. Relative-density tests cannot be performed on sealed lead–acid batteries (*i.e.* valve-regulated designs, see Section 8.7, Chapter 8) as there is no free electrolyte. Nor can they be conducted on unsealed secondary nickel batteries as the potassium hydroxide electrolyte does not participate in the cell reaction and thus remains at constant concentration.

For many applications, notably electric-vehicle traction batteries, it is highly desirable to have a direct meter-reading of the state-of-charge of the battery in order to determine the remaining operational time before recharge is required. This is the electrical equivalent of the fuel gauge in a conventional car. For a battery with a sloping discharge curve (*e.g.* lead–acid: Figure 2.5, Chapter 2) this can be a simple voltmeter, provided allowance is made for the discharge rate and the temperature. A more direct method, and the only one applicable to batteries with flat discharge curves, is to include in the circuit some form of coulometer which measures directly the charge in ampere-hours which has been taken from the battery.

Another aspect of secondary battery testing is to ascertain how much cycle-life is left in the battery. In effect, it is necessary to determine the residual capacity of the battery when fully charged. The most direct, but least convenient, method is to subject the battery to a constant-current discharge and to record the time to the cut-off voltage. This is time-consuming and may be more appropriate in the laboratory than in the working situation. More usually, a measurement is made of the 'conductance' of the battery, which is the inverse of the internal resistance. The conductance is expressed in siemens (S), where 1 S is the reciprocal of 1 ohm (Ω). Most batteries have a resistance of the order of milliohms (mΩ) and, therefore, a conductance of the order of kilosiemens. As a battery ages and its effective capacity declines, its conductance diminishes also in an approximately linear fashion. An example of a simple battery conductance meter, in both analogue and digital forms, is shown in Figure 4.8.

(a)

(b)

Figure 4.8 *Simple battery-conductance test meters in* (a) *digital form and* (b) *analogue form*
(By courtesy of Batteries International)

For large battery banks, in either stationary or traction applications, it is no longer sufficient simply to monitor the state of the battery. It is also essential to exercise some control over the charge and discharge of individual cells and modules. In a typical battery management system, multiple sensors measure the on-line voltage and temperature of each unit, and sometimes other parameters such as internal pressure. The information is recorded by multi-channel data loggers and fed into a computer which acts as the central control unit. The latter compares the acquired data with pre-set limits, with due allowance for the ambient

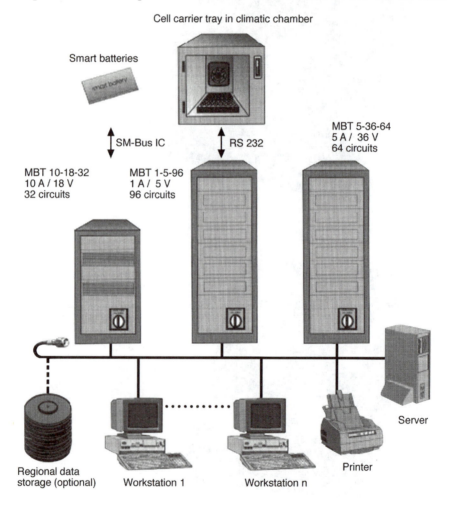

Figure 4.9   *Schematic representation of a battery management system*
            (By courtesy of UK & International Press)

temperature. When any parameter falls outside of these limits, the control unit is able to initiate corrective action, *e.g.* by reducing or cutting off the current to a module, by bypassing modules, or by switching on or off cooling fans. Actuators may be used to operate contactors or to open or close valves in water-cooling circuits.

A battery management system consists of hardware (sensors, data logger, central computer, actuators) and a versatile software program written especially for the battery type which is being controlled. A schematic representation of a battery management system is given in Figure 4.9. In this system, the battery pack is housed in an environmental chamber and there are multiple, networked workstations for monitoring the battery condition. Such battery management systems are highly adaptable and can be used for both pulsed-current and conventional charging. Simpler, on-board systems are often fitted to electric vehicles to monitor and control the charging and discharging of the traction battery. There is good evidence that such facilities prolong battery life.

Part 2

Primary Batteries

By far the majority of primary batteries are small, single cells which are used in domestic applications. These are usually of cylindrical configuration and are produced in various standard sizes. Sometimes, however, two or more cells are joined together in series to provide a higher voltage. In such arrangements, the small battery pack may have a different configuration, for example, a prismatic (rectangular) shape. Other primary cells of smaller size are manufactured in the form of button cells or coin cells to power watches, calculators, and hearing aids. Collectively, all such batteries are known as 'consumer batteries'.

Most cylindrical cells are based on the well-known zinc–manganese dioxide couple. Older (and cheaper) cells of this type are the so-called 'zinc–carbon' cells. Modern cells, of higher performance and greater cost, are known as 'alkaline' cells. These two cells types are described in Chapter 5, along with button and coin cells which retain the zinc negative electrode, but which employ silver oxide, mercury oxide, or simply air, in place of manganese dioxide as the positive electrode.

In recent years, lithium primary batteries have come into prominence. Their characteristics are described in Chapter 6. Lithium button and coin cells have replaced zinc cells for certain applications and larger cylindrical cells are now available. For specialized applications, mostly in the military and electronics fields, still larger lithium cells are manufactured. These have liquid or dissolved positive electrodes which are combined with the electrolyte.

The discussion of primary batteries concludes, in Chapter 7, with a description of specific uses for these batteries in medicine, and in the marine and the defence fields. The chapter reviews some novel types of primary battery and also examines the very particular requirements which batteries have to meet in such applications.

Chapter 5

Zinc Primary Batteries

In this chapter, we describe consumer primary batteries which use zinc as a negative electrode. Lithium primary batteries are discussed in Chapter 6 and various rechargeable batteries in Part 3.

The most common consumer batteries are based on the zinc–manganese-dioxide electrochemical couple and are manufactured as two basic types, namely, zinc–carbon and alkaline-manganese. The latter are often referred to simply as 'alkaline batteries'. Both battery systems have a nominal voltage of 1.5 V and are produced in a range of sizes. The principal manufacturers include Ralston Purina (Energizer, Ever Ready, Ucar), Duracell (Mallory), Rayovac (Vidor), Varta, SAFT, Panasonic, and Kodak. Although cell sizes are nominally standardized in terms of their physical dimensions, not all companies use the same designatory letters. Various equivalents are listed in Table 5.1, and a selection of cells offered by different manufacturers and retailers is shown in Figure 5.1.

5.1 ZINC–CARBON CELLS

The first zinc–manganese dioxide cell was introduced in the middle of the 19th century by the French scientist Georges Leclanché, as discussed in Chapter 1. It was a wet cell and, by today's standards, was very primitive in design. Nevertheless, it provided a continuous, small current at a potential of 1.5 V, and allowed the development of the telegraph service which was established during the latter stages of the 19th century. Later advances led to the so-called 'dry cell'. In this design, the liquid electrolyte was immobilized in an inert powder to form a paste and the zinc rod (the negative electrode) was replaced by a zinc can, which also acted as a container for the other cell components.

Fifty years ago, most primary batteries were of the Leclanché type. A central carbon rod, which served as the positive electrode, was sur-

Table 5.1 *Size specifications of consumer cells*

Cell size	Equivalent	Duracell	Nominal diameter (mm)	Nominal height (mm)
N		MN9100	12.00	30.20
AAA	LR03 or R03	MN2400	10.50	44.50
AA	LR6 or R6	MN1500	14.50	50.50
C	LR14 or R14	MN1400	26.20	50.00
D	LR20 or R20	MN1300	34.20	61.50
PP3	6LR61	MN1604	[a]	[a]

[a]The PP3 unit is a 9 V, prismatic (rectangular) shaped battery of nominal height 47 mm, width 26 mm, and thickness 17 mm. It contains six series-connected cells.

Figure 5.1 *Some consumer cells in popular sizes. Back row (left to right): N-size alkaline cells by Energizer; AAA-size alkaline cell by Varta; AA-size cell by Union Carbide (alkaline); zinc–carbon cell by Vidor (Rayovac); alkaline cell by Kodak; nickel–cadmium rechargeables by Boots and Ever Ready. Front row (left to right): C-size zinc–carbon cells by Ever Ready; PP3 prismatic alkaline cell (9 V) by Duracell; D-size alkaline cells by Duracell*

rounded by crude manganese dioxide (MnO_2, often the mineral pyrolusite) and carbon powder that were intimately mixed together. The function of the carbon powder was to increase the electrical conductivity of the positive active-mass, which thus reduced the internal resistance of the cell. The electrolyte, an aqueous solution of ammonium chloride (NH_4Cl) and zinc chloride ($ZnCl_2$), was absorbed into both the pores of a paste-type separator (*e.g.* gelled flour and starch layers) and the mixture of manganese dioxide and carbon. Hence, the term 'dry cell'. In reality, the cell was still a 'wet' cell with a liquid electrolyte, albeit an immobilized liquid. The cell had a seal (asphalt, wax) and a vent at its upper end, together with a cap which served to insulate the positive carbon rod from the negative zinc can. Finally, the zinc can was surrounded by a cardboard jacket, on which the manufacturer's name and other information was printed. One of the defects of this early design of cell was that the zinc can took part in the discharge reaction and, therefore, became progressively thinner until, eventually, it perforated. In effect, the cell chemicals were then contained only by the cardboard enclosure. The result was that discharged cells leaked badly and corroded the equipment in which they were housed.

Over the years, improvements have been made in the design of Leclanché cells and their materials of construction. The metallurgy of the zinc can has been improved through alloying additions to facilitate deep drawing, a process which allows cans to be produced in one piece, *i.e.* without the need for soldering or welding. Better designs of seal have been developed. The use of mercury to increase the overpotential of hydrogen evolution at the zinc electrode has been largely phased out for environmental reasons. The flour–starch paste separator has been replaced by paper which is thinly coated with the same materials. The cardboard jacket has now been substituted by a steel container coated with polymer and a polyester film label, or by a polymer jacket. When a steel outer case is used, it is insulated from the zinc can. The insulating cell cap is no longer made from cardboard, but from a hard polymer. These improvements reduce greatly the tendency of the cells to leak electrolyte when fully discharged. The construction of a modern zinc–carbon cell is shown in Figure 5.2. This design of cell is known as the 'bobbin cell'. The other principal design of consumer cell is the 'spirally-wound' or 'jellyroll' cell, which is described in Section 6.3 of Chapter 6 and shown in Figures 6.4 and 6.6.

Leclanché cells are in declining use today. Although cheap to manufacture and purchase, they suffer from a number of limitations, namely:

- the cells are not suitable for high-drain applications (such as driving electric motors) as they readily polarize and the available capacity falls sharply with increasing discharge rate;
- the shelf-life is not especially long (about two years);
- the optimum temperature range of operation is 10 to 40°C; outside of this range, the performance deteriorates markedly.

The cells are best suited to low-drain, intermittent applications, with rest periods for recuperation (depolarization) to take place, and for use within

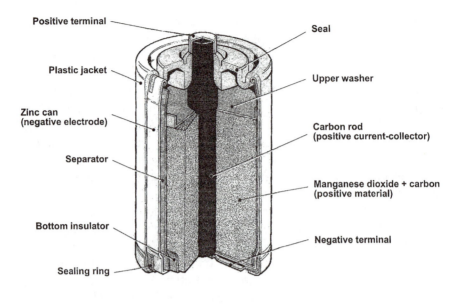

Figure 5.2 *Construction of a modern zinc–carbon cell*
(By courtesy of Duracell Batteries)

about two years. A good example is a domestic flashlamp battery. Other applications include clocks, portable radios, door-chimes and smoke detectors, but the relatively short operational life of the batteries is a nuisance in that periodic replacement is necessary.

Superficially, the electrochemistry of the Leclanché cell is simple. At the positive electrode, manganese dioxide is reduced to a trivalent manganese compound with a resultant rise in pH, *i.e.*

$$MnO_2 + H_2O + e^- \longrightarrow MnO \cdot OH + OH^- \tag{5.1}$$

At the negative electrode, zinc metal is oxidized to Zn^{2+} ions, *i.e.*

$$Zn \longrightarrow Zn^{2+} + 2e^- \tag{5.2}$$

The pH in the vicinity of this electrode decreases due to hydrolysis of the Zn^{2+} ions *via* reactions such as:

$$Zn^{2+} + H_2O \longrightarrow Zn(OH)^+ + H^+ \tag{5.3}$$

In reality, the cell reactions are much more complex than this, with the formation of sparingly soluble intermediates such as $Zn(NH_3)_2Cl_2$, $ZnO \cdot Mn_2O_3$ and $ZnCl_2 \cdot 4Zn(OH)_2$, particularly at high discharge rates.

These solids tend to impede the ionic diffusion which is necessary for cell discharge. The resulting fall in pH near the zinc electrode leads to enhanced chemical corrosion, with liberation of hydrogen. As the internal pressure in the cell rises, the seals ultimately leak and force electrolyte out of the cell.

Many of these problems are overcome in the 'zinc chloride' cell which has a simpler chemistry because the ammonium chloride electrolyte is replaced entirely by $ZnCl_2$ solution. The reduced tendency for the electrodes to be blocked by solid products permits faster diffusion and higher rates of discharge. Manganese dioxide of better quality, prepared electrolytically, is employed. A greater proportion of carbon black is included to facilitate the discharge reaction at the positive electrode by improving the electrical conductivity of the mix. The cells also have a more sophisticated seal design. These factors, taken together, result in a higher cost of manufacture, which offsets the improved performance. In fact, zinc chloride cells fall between Leclanché and alkaline-manganese cells (see next section) with respect to both performance and cost. They perform better than Leclanché cells in high-drain duty and, particularly, when used at moderately low temperatures (to $-10°C$). The cells are offered for sale under various trade names such as 'High Power', 'High Performance', 'Heavy Duty', 'Grade 1', and 'Silver Seal', as well as simply 'Zinc Chloride', and are capturing an increasing share of the zinc–carbon battery market.

Invariably, the user has no knowledge of, or little interest in, the actual current flowing in a consumer cell. For this reason, as discussed in Chapter 2 (see Figure 2.7), discharge curves are not normally plotted as voltage against capacity at constant current, as for larger secondary batteries, but as voltage against hours of service through a given load. As discharge progresses, the internal resistance of the cell increases and, correspondingly, the current falls. A comparison of the performance of Leclanché and zinc chloride cells (both D-size) discharged continuously at a high rate through a $3.9\,\Omega$ resistor is presented in Figure 5.3. It is evident that the 'High Power' cell gives a far longer service-life to the cut-off voltage of 0.9 V. The difference in performance is less marked at lower currents, especially if the discharge is discontinuous with rest periods interposed.

D-size and AA-size zinc chloride cells discharged through the same $3.9\,\Omega$ resistor deliver 10 and 1.2 h of service, respectively. This difference is mostly a matter of size and the amount of chemically reactive materials present, although proportionately the larger (D-size) cell gives a longer service-life as its current density is lower.

The effective capacity available from either cell depends critically on

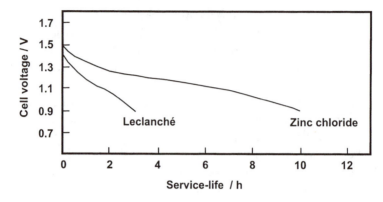

Figure 5.3   *Discharge curves for Leclanché and zinc chloride ('High Power') cells (D-size; at 20°C through a 3.9Ω resistor)*
(By courtesy of Duracell Batteries)

the current drawn. If the load (*i.e.* the circuit resistance) is increased by a factor of 10, thereby reducing the current dramatically, the available capacity (mAh) will increase two- to three-fold, and the service-life (at the lower current drain) about twenty-fold. Furthermore, if the discharge is intermittent rather than continuous, so as to allow time for recovery by diffusion processes in the active materials, the performance will be even better. For instance, the D-size zinc chloride cell discharged through 3.9 Ω for 1 h per day will give a service-life in excess of 16 h, compared with 10 h for continuous discharge.

The improved performance of zinc chloride cells over that of Leclanché cells is quite striking. Table 5.2 provides published values for the capacity, expressed in mAh, measured at 20°C to a cut-off of 0.9 V when the cells are discharged continuously through the stated resistance. It should be noted that the ratio of the capacities is not as great as the ratio of the service-lives shown in Figure 5.3 (*i.e.* > 3) as here the comparison is being made at higher circuit resistances and lower currents. Similarly, the improved performance at low temperatures is quite remarkable. The effect of temperature on the output of D-size cells when discharged at 0.3 A for 30 min per day is presented in Figure 5.4. At 0°C, the capacity of the zinc chloride cell is 5 Ah compared with 2 Ah for the zinc–carbon cell.

In summary, zinc chloride cells are substantially better than Leclanché cells for continuous drain applications, particularly at moderate-to-high currents and/or at low temperatures. Leclanché cells perform satisfactorily for intermittent operation, with extended recuperation periods between discharges, if the current is not too high and the temperature is not too low.

Table 5.2 *Capacity of zinc chloride and Leclanché cells discharged continuously through the stated resistance*

Cell size	Resistance (Ω)	Capacity (mAh)	
		Zinc chloride	Leclanché
AA	75	1100	835
C	39	3300	1750
D	39	7300	4700

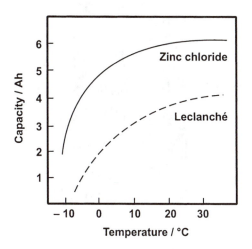

Figure 5.4 *Low-temperature performance of Leclanché and zinc chloride cells (D-size)* (By courtesy of Ever Ready)

5.2   ALKALINE-MANGANESE CELLS

Alkaline-manganese, the premium form of the $Zn–MnO_2$ cell, differs from zinc–carbon in a number of important respects. A principal difference lies in the use of concentrated (~ 30 wt.%) potassium hydroxide as the electrolyte solution. This electrolyte is chosen primarily on account of its high electrical conductance (Table 5.3). Another significant difference is in the nature of the negative electrode which consists of finely-divided zinc powder (or zinc granules) packed around a current-collector positioned at the centre of the cell. A third difference lies in the cell configuration; the positive-electrode mix, of electrolytic MnO_2 and fine graphite powder, is packed around the *outside* of the zinc negative and separator, and is in electrical contact with a nickel-plated steel can (Figure 5.5). This

Table 5.3 *Electrical conductance of 30 wt.% KOH solution*

Temperature (°C)	40	20	0	− 20
Specific conductance (S cm^{-1})	0.75	0.53	0.35	0.17

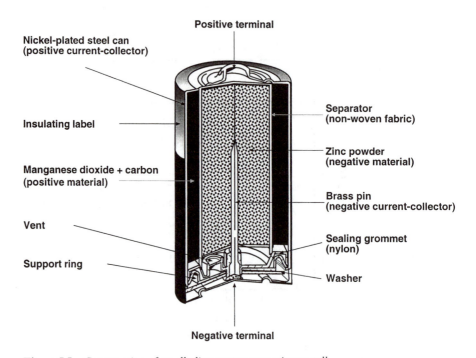

Figure 5.5   *Construction of an alkaline-manganese primary cell*
              (By courtesy of Duracell Batteries)

inversion of the cell configuration might be expected to lead to an inversion of polarity of the terminals. As this would be unacceptably confusing to the consumer, the problem is overcome by bringing the central negative current-collector, in the form of a brass pin, into contact with the cell base (rather than the cap) and having the insulating seal at the *bottom* of the cell rather than the top, with the steel can contacting the top (positive) terminal.

The chemical reactions taking place in an alkaline cell are intrinsically simpler than those in a Leclanché cell as there is no ammonium chloride or zinc chloride present to form blocking precipitates. The overall reaction is simply:

$$Zn + 2MnO_2 + H_2O \longrightarrow ZnO + 2MnO \cdot OH \qquad (5.4)$$

ough with $Zn(OH)_4^{2-}$ ions formed as intermediate species. There is no change in the concentration of OH^- ions during discharge. Neverthe-s, the internal resistance of the cell increases steadily as a result of other tors. Typically, the resistance of an AA-size cell will rise from 150 mΩ en fully charged to 600 mΩ when discharged.

Because concentrated potassium hydroxide electrolyte is chemically ctive, and is also prone to flow (creep) along surfaces, the development satisfactory seals for the alkaline-manganese system was a technical llenge which had to be solved before the cells could be commercial-d. Modern cells have effective seals which are generally resistant to kage and corrosion. The seals are made from a plastic material such as on or polypropylene.

Alkaline-manganese cells, which retail at a premium price, are more table than zinc–carbon cells for continuous, high-drain duty and for at low temperatures. They also have a longer shelf-life (more than 4 rs at 20°C) and are thus attractive for low-drain, 'fit and forget', blications such as smoke detectors. Their principal economic advan-e is seen, however, in high-drain applications such as children's toys, sette players and electric shavers, where the effective capacity de-ered is many times that of a zinc–carbon cell under similar duty. A nparison of the approximate service-life of alkaline-manganese and c–carbon cells at various discharge rates is given in Table 5.4. It will be n that the ratio of hours of service (alkaline-manganese *vs.* zinc–car-a) is greater for the large D-size cells than for the smaller AA-size cells, d is also greater at high discharge rates (small loads). At lower dis-rge rates than shown, the ratio falls to a value of about two at 20°C, rises again sharply at temperatures below 0°C where alkaline cells are ch superior, even at low discharge rates.

The superior performance of alkaline cells at low temperatures is

Table 5.4 *Approximate service-lives (in hours) of alkaline-manganese (AM) and zinc–carbon (Zn–C) cells at different continuous discharge rates (data published by Duracell for discharge to 0.8 V at 21°C)*

Load (Ω)	D-size cells			AA-size cells		
	AM (h)	Zn–C (h)	Ratio	AM (h)	Zn–C (h)	Ratio
3	45	3	15	5	0.8	6.3
5	80	6.5	12	10	1.5	6.7
10	130	13	10	15	3.3	4.5
30	450	80	5.6	65	13	5
50	700	150	4.7	120	30	4

illustrated in Figure 5.6 which shows the influence of temperature on the discharge of a D-size cell through a 3.9 Ω resistor. Although the performance falls off drastically with decreasing temperature, the cell is still operable for several hours at −30°C, which is in marked contrast to zinc–carbon.

Recently, at least two companies have introduced a premium grade of alkaline battery for high-drain applications. This has been made possible by advances in materials science which have reduced the internal resistance of the cell. In effect, the resistance has been lowered by applying coatings to the positive and negative current-collectors, by using a finer grade of graphite powder, and by packing more MnO_2 powder into the space available for the positive electrode. Coatings will prevent, or reduce, the build-up of corrosion products on the current-collectors, while a finer grade of graphite will improve electrical point contacts and electronic conduction in the positive-electrode mix.

The claims for improvements in discharge power and service-life are quite impressive. The energy available from Duracell 'Ultra' and standard alkaline AA-size cells is plotted against the continuous power output on a log–log scale in Figure 5.7. The advantage of the 'Ultra' cell is most pronounced at high current drains. At a continuous power output of 0.5 W, the energy available from the 'Ultra' cell is about 30% greater than that from the standard alkaline cell, and rises to about 100% greater energy at an output of 1 W. On the other hand, as Figure 5.8 shows, the service-life to a cut-off voltage of 0.9 V at an output of 1 W is less than one-third that at 0.5 W and, indeed, is less than 1 h. This is a high discharge rate for an AA-size cell and under such conditions a rechargeable cell may well be preferred. Nevertheless, the higher voltage and

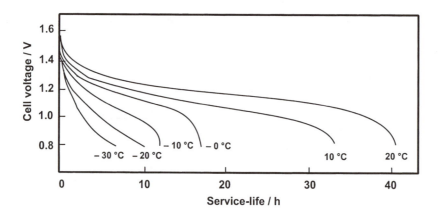

Figure 5.6   *Low-temperature discharge of a D-size alkaline cell through a 3.9 Ω resistor* (By courtesy of Duracell Batteries)

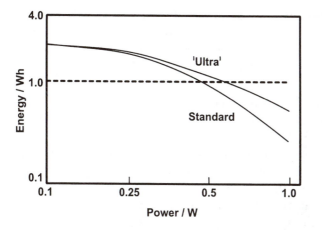

Figure 5.7 *Energy available from Duracell 'Ultra' and standard alkaline cells (AA-size) as a function of continuous power output (log–log plot). Delivered energy to a cut-off voltage of 1.0 V*
(By courtesy of Duracell Batteries)

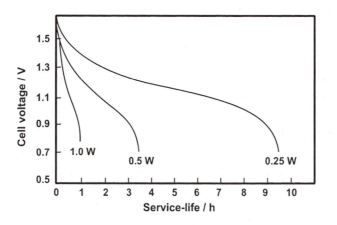

Figure 5.8 *Discharge curves for Duracell 'Ultra' cells (AA-size). Discharge to a cut-off voltage of 0.9 V*
(By courtesy of Duracell Batteries)

capacity of alkaline-manganese cells compared with nickel–cadmium is a very positive feature and these advanced cells are recommended for high-power devices such as flash cameras, digital cameras and halogen flashlights where a significantly longer life is obtained.

Alkaline-manganese cells are manufactured in the same sizes and configurations as zinc–carbon cells. The most popular sizes are cylindrical AAA, AA, C and D cells, although both smaller and larger cells are made. Another popular item is the prismatic PP3 battery; this has a voltage of 9 V and is ideal for low-drain applications such as smoke detectors. The typical rated capacity of cylindrical alkaline cells of size AAA, AA, C and D is 1.18, 2.7, 7.75 and 18 Ah, respectively. It should be noted, however, that these values are strongly dependent upon the discharge rate. The corresponding approximate service-life of the same cells, when discharged intermittently at room temperature through a $10\,\Omega$ resistor to 0.8 V, is 9, 21, 40 and 140 h, respectively.

Some alkaline cells are provided with an external indicator strip, designed to show the approximate state-of-charge of the cell. This strip incorporates a fine carbon thread which connects the positive and negative terminals when two markers in the external wrapper are depressed to complete the circuit. The thread warms up, according to the current passed, and activates a heat-sensitive dye included in the strip. The colour of the dye gives an indication of the local temperature in the strip and, therefore, of the current flowing and, in turn, the cell voltage. It has to be said, however, that the colour change is fairly subtle and a simple domestic test meter provides a more accurate measure of the state-of-charge of the cell.

It is interesting to reflect on the cost of electricity from primary batteries. A blister pack of two D-size alkaline batteries retails in the UK for around US$4.50. These have a combined rated capacity of 36 Ah, equivalent to 54 Wh. On this basis, the cost of electricity is about US$83 per kWh. By comparison, 1 kWh of mains electricity costs only US$0.10, *i.e.* close to a thousand times cheaper!

5.3   BUTTON AND COIN CELLS

Manganese dioxide is by no means the only oxidant which may be used in zinc-alkaline cells. Other possibilities are silver oxide, mercury oxide, and oxygen (air). All of these have been developed as practical cells, but today their consumer use is largely confined to button and coin cells, so-called 'miniature cells', as found in electronic circuitry. Some larger cells are employed in specialized applications, particularly in the defence and aerospace fields. Button cells are taller and usually smaller in diameter than coin cells. Button cells are used in small alarm clocks, timers, and often in watches. Coin cells are preferred for pocket calculators because they need to be flat, while still of greater capacity than watch batteries. The choice of button and coin cells available to design engineers is very great indeed. For instance, Energizer manufacture no fewer than 34 different cells based on the zinc–silver-oxide system for use in watches and travel clocks. These cells range in capacity from 5.5 to 181 mAh.

Other companies manufacture to the same standard sizes.

The construction of a typical zinc–silver-oxide or zinc–mercury-oxide button cell is shown schematically in Figure 5.9. It should be noted that the base and outer can of a button or coin cell is the positive terminal, while the cap is the negative.

5.3.1   Zinc–Silver-oxide Cells

The zinc–silver-oxide cell was invented in the middle of the 19th century, but was not truly practicable until the development of ion-permeable membranes as separators at a much later date. The cell was introduced commercially by Union Carbide in 1961 to satisfy the new market of electric watches. The electrolyte is a concentrated solution of potassium hydroxide (for high-rate duty) or sodium hydroxide (for long life). The overall cell reaction is:

$$Zn + Ag_2O \longrightarrow ZnO + 2Ag \qquad (5.5)$$

The open-circuit voltage is 1.60 V.

Different applications for these cells demand a wide range of discharge currents: from a few microamperes for watches with liquid crystal diode (LCD) displays, to tens of milliamperes for light-emitting diode (LED) displays or for audible alarms. As well as their use in watches, zinc–silver-oxide batteries find prime use in pocket calculators and in electronic instruments. They are characterized by having a very flat discharge voltage profile between 1.50 and 1.55 V, which is ideal for watches (Figure

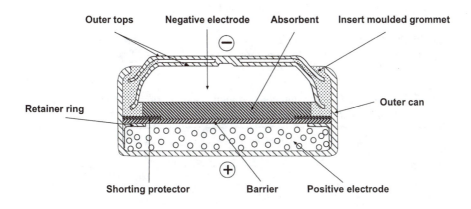

Figure 5.9   *Schematic representation of the construction of a typical button cell (By courtesy of Academic Press)*

5.10). It is noteworthy that the load voltage is almost independent of the current drain. This indicates a low cell resistance and remarkably little polarization.

Zinc–silver-oxide cells may be used down to 0 °C at a current drain of 1 mA, and to − 10 °C at 0.1 mA. They have a high volumetric capacity (200 to 300 mAh cm^{-3}), *i.e.* up to 50% better than that of either alkaline-manganese or lithium miniature cells (see Table 5.5). This is an important parameter for a watch battery. The storage life of zinc–silver-oxide cells is also excellent, *viz.* 80% retention of capacity after four years at 20 °C. Long-life applications include camera memory functions, remote access car keys and, as mentioned above, pocket calculators.

5.3.2 Zinc–Mercury-oxide Cells

These cells are closely analogous to zinc–silver-oxide cells, but have a somewhat lower open-circuit voltage (1.35 V). A small proportion of

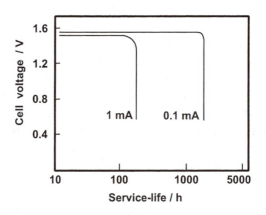

Figure 5.10 *Discharge curves for zinc–silver-oxide button cell (capacity: 200 mAh); low-rate (0.1 mA) and medium-rate (1 mA) discharge at 20 °C*
(By courtesy of Ever Ready)

Table 5.5 *Gravimetric and volumetric capacity of miniature primary cells (data from manufacturers' literature)*

Battery	Voltage (V)	Gravimetric capacity (mAh g^{-1})		Volumetric capacity (mAh cm^{-3})	
		Duracell	*Energizer*	*Duracell*	*Energizer*
Alkaline	1.5	35–70	—	130–230	—
Zn–AgO	1.5	55–75	—	220–300	190–310
Zn–air	1.4	225–320	220–330	770–1070	580–1050
Lithium coin	3.0	40–80	30–70	140–200	160–220

manganese dioxide is sometimes added to the positive electrode mix to raise the cell voltage to between 1.40 and 1.45 V. The pellet of mercuric oxide (HgO) has finely-divided graphite added to increase the electrical conductivity and to minimize the coalescence of liquid mercury droplets formed by the discharge reaction. Mercury oxide cells have flat discharge curves similar to those of silver oxide cells, but even lower internal cell resistance. These features make the mercury system suitable for applications where constant voltage during discharge is important or relatively high currents are required. The cells have practical volumetric capacities which approach those of zinc–silver-oxide counterparts. Nowadays, they are not so extensively manufactured as zinc–silver-oxide cells and are being phased out, in part because of environmental considerations associated with mercury disposal.

5.3.3   Zinc–Air Cells

The air electrode is well known in alkaline fuel cells and there is no reason in principle for not using it in zinc primary batteries. Zinc–air button cells are similar in appearance to zinc–silver-oxide cells, except for a small hole/holes in the positive base plate of the can to admit air to the electrode. When supplied, the holes are covered with adhesive tape which must be removed before the cells are put into operation. The purpose of this tape is to exclude carbon dioxide during storage that, otherwise, would react with the potassium hydroxide electrolyte to produce potassium carbonate and, thereby, reduce the conductivity of the electrolyte.

The electroreduction of oxygen requires a catalyst and may be represented by:

$$O_2 + H_2O + 2e^- \longrightarrow HO_2^- + OH^- \tag{5.6}$$

$$HO_2^- + H_2O + 2e^- \longrightarrow 3OH^- \tag{5.7}$$

where HO_2^- is the intermediate hydroperoxide ion. The catalyst generally employed in alkaline media is nickel, often deposited on a porous carbon substrate. As this electroreduction occurs only at a three-point contact between air, liquid electrolyte and solid catalyst (except to the extent that the oxygen may dissolve in the electrolyte), it is necessary to arrange that the interface between gas and liquid is stable within the pores of the catalyst. Thus, electrolyte must be absorbed into the pores of the catalyst, but must not flood it to the exclusion of gas. This requires the careful construction of the catalyst layer. On the electrolyte side, the catalyst layer must be made hydrophilic, but on the gas side it needs to be hydrophobic, although porous to gas. To achieve this, the surface exposed to air is treated with a water-repellent substance, such as paraffin wax, which coats the walls of the pores on that side of the catalyst layer. The electrolyte | gas interface is thereby stabilized within the catalyst layer.

As the air electrode is only a catalyst and not the active material itself, it is much thinner and lighter than the silver oxide or mercury oxide electrodes in the other button cells. This allows space to accommodate considerably more zinc in the zinc–air cell, which makes for extremely

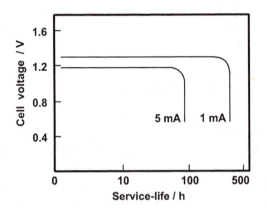

Figure 5.11   *Discharge curves for zinc–air button cells*
(By courtesy of Ever Ready)

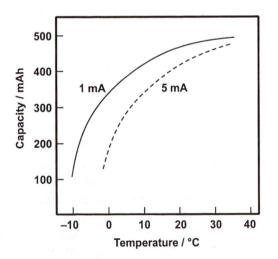

Figure 5.12   *Effect of temperature on the capacity of zinc–air button cells*
(By courtesy of Ever Ready)

Figure 5.13   *An 8.4 V battery and button cells of the zinc–air system*
(By courtesy of Duracell Batteries)

high gravimetric and volumetric capacities (Table 5.5). For example, the capacity (and therefore the service-life) of a zinc–air battery designed for hearing aids is more than twice that of the equivalent zinc–mercury-oxide cell. The open-circuit voltage of the zinc–air cell is 1.4 V and the discharge curve is flat (Figure 5.11). The current drain for zinc–air button cells lies in the range 1 to 10 mA and the cell may be used down to $-10\,^\circ$C (Figure 5.12). Because air is excluded from the cell until immediately prior to use, the capacity loss on storage is low compared with other button cells.

The zinc–air button cell has found its widest application as the power source for hearing aids. Typical capacities lie in the range 100 to 500 mAh. The cell is ideal for low-drain appliances which are in frequent or continuous use. Once activated, however, its useful life is just a week or two as the electrolyte is slowly poisoned by carbon dioxide. For this reason, cells for hearing aids are sold in blister packs of 6 to 8 units. A selection of zinc–air button cells and a 8.4 V (6-cell) battery, as manufactured by Duracell, are shown in Figure 5.13.

An Israeli company, Electric Fuel Corporation, is now promoting small zinc–air primary cells for use in mobile telephones. These cells provide three to five times more 'talk-time' than rechargeables, which is useful when distant from a source of electricity for recharging. Also, being primaries, the zinc–air cells are used straight from the packet without the need for a preliminary charge.

Chapter 6

Lithium Primary Batteries

Over the past decade or two there has been a growing interest in lithium primary batteries. Their impact on the consumer market has been mostly as button and coin cells for electronics applications. Larger cells (AA-, C- and D-sizes) have been manufactured principally for military use, but gradually these are being adopted for civil applications as well.

Lithium has two unique properties which make it highly suitable as a negative electrode for batteries. First, it is the lightest metal in the Periodic Table, with an atomic mass of only 6.94 (*cf.* zinc 65.37, cadmium 112.40, lead 207.19). This is a most attractive property for a lightweight battery. The specific capacity of lithium metal is 3.86 Ah g^{-1}, compared with 0.82 Ah g^{-1} for zinc. Second, lithium has a high electrochemical reduction potential, *viz.* (3.045 V, which holds out the promise of a 3 V battery when combined with a suitable positive electrode. The conjunction of the two properties should result in a cell of high specific energy. On the debit side, the electrooxidation of lithium is a one-electron process, whereas that of zinc and cadmium are two-electron processes. It follows that lithium cells will not have so great an advantage over zinc or cadmium cells in volumetric terms (Wh dm^{-3}) as in gravimetric terms (Wh kg^{-1}).

The major problem with lithium is that it is highly reactive towards water and, therefore, cannot be used with an aqueous electrolyte. What are the alternatives? In principle, there are five classes of non-aqueous electrolyte that can be used in lithium batteries. These are: (i) solutions of lithium salts in polar organic liquids (note, a polar organic molecule is one which has a significant dipole moment as a result of having an asymmetric structure); (ii) solutions similar to those in (i) but in polar inorganic liquids; (iii) fused lithium salts; (iv) ionically conducting polymers; (v) ionically conducting ceramics. Most lithium cells and batteries employ the first of these possible electrolytes, although a few use polar inorganic solvents, while some high-temperature military batteries employ fused salts (see Section 7.4, Chapter 7). Polymers which are adequately conducting to lithium ions have been developed recently and lithium–polymer batteries are now coming on to the market. No lithium-conducting ceramic which is satisfactory for use in batteries has been identified. This contrasts with the situation for sodium, where sodium beta alumina has been widely employed as an electrolyte in sodium batteries (see Section 11.5, Chapter 11).

6.1 LITHIUM CELL CHEMISTRY

Thermodynamically, lithium is unstable with respect to most organic liquids, just as it is with respect to water. The factor which allows an organic liquid to be used as electrolyte is the formation of a stable passivating layer on the surface of lithium metal. (This may be seen as analogous to the stabilizing oxide film on stainless-steel which prevents the steel from rusting.) The passivating layer formed on exposure of lithium metal to air is very thin and is a mixture of lithium nitride (Li_3N) and lithium hydroxide/carbonate. The conductivity for lithium ions is sufficiently high for the layer to be regarded as a second, solid electrolyte additional to the main liquid electrolyte. The layer does not normally contribute substantially to the internal resistance of the cell, but does protect the lithium from being self-discharged on open-circuit stand, which would otherwise occur as a result of corrosion processes. The choice of organic electrolyte for lithium cells is determined by the detailed chemistry of the interface reaction and by the stability and properties of the resultant surface film. In particular, the passivating film should cause minimal voltage delay at the start of discharge, and should reform rapidly after a high-current discharge pulse.

The second factor which is important in the choice of an organic electrolyte is the ionic conductivity of the liquid as this determines the internal resistance of the cell. At best, the conductivity will be markedly inferior to that of usual aqueous electrolytes (zinc chloride, potassium hydroxide, sulfuric acid, *etc*.). Thus, it is important to maximize the conductivity as far as possible. This suggests the use of a polar solvent which is likely to facilitate the dissolution of lithium salts as ionic species. Among those which have been employed are: linear esters (methyl formate, methyl acetate, diethyl carbonate), cyclic esters (ethylene carbonate, propylene carbonate), linear ethers (dimethoxyethane), cyclic ethers (dioxolane). The selection of the best solvent is influenced by the extent to which it can dissolve the chosen lithium salt, as this ability largely determines the conductivity of the solution.

Other properties of the solvent to be considered are its melting point and boiling point (which may limit the temperature range over which the battery can operate), its density and viscosity (which help to determine the conductivity), and its chemical compatibility with the positive electrode. Among the dissolved salts which have been investigated thoroughly are $LiClO_4$, $LiAlCl_4$, $LiBF_4$, $LiPF_6$, $LiAsF_6$, $LiCF_3SO_3$, and $LiN(CF_3SO_2)_2$. Solutions of these salts in the above solvents provide reasonable conductivities ($\sim 10^{-2}$ S cm^{-1}). Although there is much information in the scientific literature on the properties of these solutions, individual battery manufacturers do not always divulge their choice of electrolyte. There seems to be considerable interest in $LiPF_6$ and $LiAsF_6$ as solutes and dimethoxyethane or diethyl carbonate as solvent.

The relatively low conductivity of the organic-based electrolytes tends to limit the power output of lithium cells which may be seen as devices with high specific energy but relatively low power. On the other hand, the low freezing points of these solutions permit the cells to operate at lower temperatures than aqueous electrolyte cells.

Purification of the electrolyte solutions is most important, particularly with respect to removing the last traces of water. As noted above, lithium metal is highly susceptible to attack by moisture and thus it must be handled, and the cells assembled, in a dedicated, low-humidity, dry room. This is a significant factor in the cost of manufacturing lithium cells.

The choice of positive electrode material for use in lithium cells is also quite wide. Among the materials which have been investigated are CuO, CuS, CF_x, FeS_2, MnO_2, MoO_3, Ag_2CrO_4, and V_2O_5. The most widely adopted for primary cells are MnO_2 and CF_x (a form of fluorinated graphite), which both give 3 V cells, and FeS_2 and CuO, which give 1.5 V cells. The latter two cells are interchangeable with alkaline-manganese cells.

6.2   BUTTON AND COIN CELLS

Primary lithium cells have a number of characteristics that make them ideal for applications which require coin and button designs, *e.g.* watches, cameras and calculators. Beneficial features include the following.

- *High cell voltage.* Lithium cells commonly have voltages of 3 V and over, as determined by the choice of positive electrode material.
- *Flat discharge.* Lithium cells exhibit flat discharge curves right up to the end of discharge. This is particularly important for use in watches and calculators.
- *Long shelf-life.* Storage lives of at least 10 years at room temperature

are possible. This is attributable to the stability of the passivating layer which is formed on the surface of the lithium metal.

- *Wide range of operating temperature.* Lithium cells may be operated from -30 to $+60°C$.

Manganese dioxide is the most usual material for positive electrodes in lithium coin and button cells. The open-circuit voltage of such cells is 3.2 V and the overall discharge reaction is:

$$Li + MnO_2 \longrightarrow LiMnO_2 \tag{6.1}$$

Because of the comparatively high resistivity of organic electrolytes compared with, for example, concentrated potassium hydroxide solution used in alkaline-manganese cells, lithium–manganese-dioxide cells are best designed with electrodes of large area and with small inter-electrode spacing for the electrolyte. These requirements favour the coin design of cell over the button cell. For the same reason, lithium coin cells are normally operated at lower currents than zinc–silver-oxide equivalents so as to minimize the resistive drop and heating in the cell.

Typical discharge curves for a lithium–manganese-dioxide coin cell are shown in Figure 6.1. Comparison with data for a zinc–silver-oxide button cell (see Figure 5.10, Chapter 5) shows that the currents are an order of magnitude lower. Size-for-size, lithium–manganese-dioxide miniature cells have a lower capacity (mAh) than zinc–silver-oxide cells (see Table 5.5, Chapter 5), but this is compensated by the doubled voltage so that the two types of cell are comparable in terms of energy density $(mWh\ cm^{-3})$.

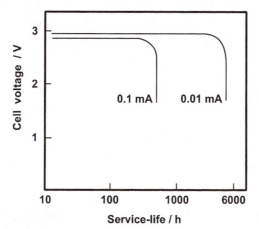

Figure 6.1 *Discharge curves for a lithium–manganese-dioxide coin cell; low-rate (0.01 mA) and medium-rate (0.1 mA) discharge*
(By courtesy of Ever Ready)

   Lithium coin cells are ideally suited to systems where a 3 V output is useful and where small continuous currents are needed for long periods, *e.g.* in memory back-up displays for computers, televisions, video cassette recorders, and cameras. At appropriately low-current drains, working lives in excess of five years may be expected. They are also useful for low-temperature applications (to $-20\,^{\circ}C$, or lower), and in situations where cell thickness is a critical factor. The configuration of a typical lithium coin cell is shown schematically in Figure 6.2. The design of the polypropylene closure is critical. It must ensure that there is no ingress of water vapour over long periods of storage or use. A selection of lithium coin cells (and cylindrical cells, see next section) manufactured by the Sanyo Corporation is shown in Figure 6.3.

6.3   CYLINDRICAL AND PRISMATIC CELLS

Cylindrical lithium cells are available with 'dissolved' or 'inorganic liquid' positive electrodes (see next section), or with conventional solid positives. The best known of the solid positive electrodes are FeS_2 (operating voltage: 1.5 V) and MnO_2 (3 V on load). Energizer manufactures AA-size lithium cells with FeS_2 as a direct replacement for alkaline cells; a comparison of the two types of cell is given in Table 6.1. Clearly, the lithium cell is capable of a significantly longer operational life, especially at high power levels (up to 1400 mA). This is important in applications demanding power pulses, such as camera flash-guns.
   How can lithium cells provide high power when the resistivity of the organic electrolyte is so much higher than that of potassium hydroxide? The answer lies in the design of the cell internals. The cell is constructed not on the bobbin principle of the alkaline-manganese cell, but on the spirally-wound ('jellyroll') principle (Figure 6.4). The lithium-foil negative

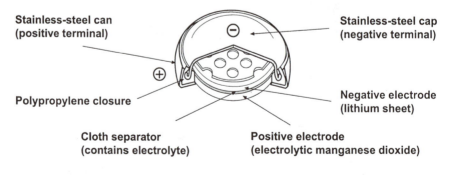

Stainless-steel can
(positive terminal)

Stainless-steel cap
(negative terminal)

Polypropylene closure

Negative electrode
(lithium sheet)

Cloth separator
(contains electrolyte)

Positive electrode
(electrolytic manganese dioxide)

Figure 6.2   *Design of lithium–manganese-dioxide coin cell*
                  (By courtesy of Ever Ready)

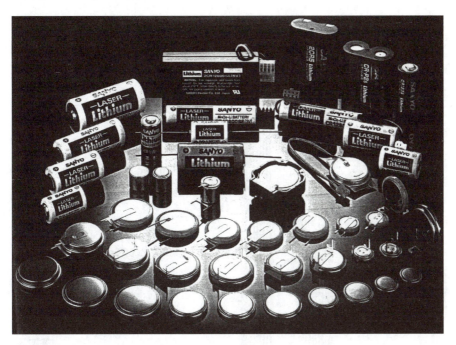

Figure 6.3 *Selection of lithium cells manufactured by the Sanyo Corporation* [By courtesy of Sanyo Energy (UK) Co. Ltd]

Table 6.1 *Comparison of lithium–iron-sulfide and alkaline-manganese AA-size cells (data taken from Energizer brochure)*

	$Li–FeS_2$	$Zn–MnO_2$
Battery mass (g)	14.5	23.0
Voltage:		
open-circuit (V)	1.8	1.6
nominal (V)	1.5	1.5
Operating life (hours to 0.9 V) at:		
1400 mA	1.3	0.2
1000 mA	2.1	0.4
400 mA	5.7	2.7
20 mA	122.0	117.0
Shelf-life (years)	10	5

electrode, the separator and the FeS_2 positive electrode are all fabricated as thin sheets and laid one upon the other to form a sandwich. This is then rolled tightly into a spiral and slid into the outer steel can. The lithium foil makes electrical contact with the nickel-plated cell base (negative), while the FeS_2 positive electrode contacts the cell cap, which is also nickel-plated. With this design, there is a large area of electrode and a very small gap between the positive and negative electrodes, so that the cell resis-

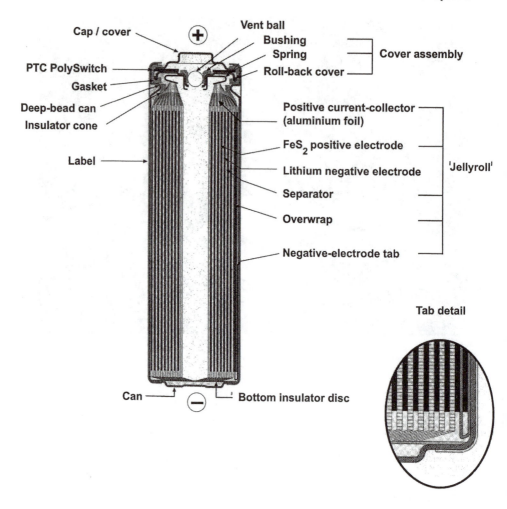

Figure 6.4 *Cross-section of lithium–iron-sulfide cell in a spirally-wound ('jellyroll') con-*
figuration
(By courtesy of Energizer UK Company)

tance (as distinct from the electrolyte resistivity) is similar to that of the
alkaline-manganese cell. The advantages of lithium–iron-sulfide cells are:

- longer service-life, especially in moderate to high-drain applications;
- flatter discharge curves than other AA-size cells;
- wide range of operating temperature (-40 to $+60\,^\circ$C);
- long storage life (up to 10 years);
- seals with good leakage resistance;
- one-third lighter in weight than conventional AA-size cells.

This is a significant list of advantages, one or more of which may appeal in many applications.

The lithium–iron-sulfide cell contains a thermal safety switch which acts as a current limiter if the cell overheats. [The resistance of the switch rises abruptly above a certain temperature and so reduces the current. Hence, the device is known as a 'PTC (positive temperature coefficient) switch'.] The service-life of the cell for continuous operation as a function of current drain and temperature is shown in Figure 6.5. The cells are advertised as being superior to alkaline-manganese for most consumer applications where AA-size cells are employed. For use in camcorders, portable computers, cellular telephones and cameras, the service-life is said to be three-to-four times that of alkaline-manganese. Naturally, the lithium cells cost more.

Lithium–manganese-dioxide cylindrical cells are manufactured primarily for professional use by electronics engineers and many of them are not in standard consumer sizes. In the spirally-wound configuration, the capacities range up to a maximum of about 1400 mAh. The cells are ideal for devices which require low background currents with short pulses of up to several amps on demand. For example, the cells are used for the rapid charging of camera flash-guns. Other applications include military communications equipment, rescue beacons, meter reading equipment, and pipeline inspection systems. Like all lithium systems, the cells have a long shelf-life and operate over a wide range of temperatures (-30 to $+60°C$).

Some companies also manufacture cylindrical lithium–manganese-dioxide cells in the bobbin configuration with a central negative electrode. These have high capacity, but deliver low discharge currents. For example, the Sanyo Corporation produces such cells with capacities up to 5 Ah, but with a standard discharge current of only 1 mA and a maximum continuous current (with loss of half the nominal capacity) of 10 mA. A variety of lithium cylindrical cells manufactured by the Sanyo Corporation is shown in Figure 6.3.

Another company, Ultralife Batteries, Inc., produces a 3 V cell of 3.6 Ah capacity and a 9 V cell of 1.2 Ah capacity, both in the prismatic PP3 configuration. These have high gravimetric and volumetric energy densities (~ 300 Wh kg^{-1} and ~ 470 Wh dm^{-3}), and also provide a continuous current drain of 300 and 120 mA, respectively. As well as having the usual advantages of lithium systems, the cells are prismatic in shape and thus may be joined in series or parallel arrays to form very volume-efficient battery packs. The same company also manufactures a range of high-rate cylindrical cells. These are spirally-wound (Figure 6.6) so as to permit high-rate discharge. The C and D-size cells have nominal capaci-

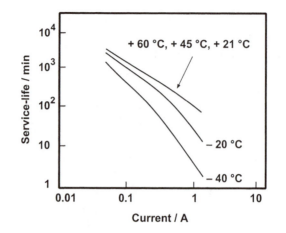

Figure 6.5 *Continuous performance characteristic of a lithium–iron-sulfide cell to 0.9 V as a function of temperature*
(By courtesy of Energizer UK Company)

ties of 4.5 and 10 Ah, and maximum continuous currents of 1.3 and 3 A, respectively. Pulsed currents may be up to 9 A (C-size) and 18 A (D-size). Such performance makes these cells ideal for many applications which require high-power pulses. They also have high energy densities (in excess of 500 Wh dm^{-3}) which are somewhat greater than those of lithium–sulfur-dioxide cells of similar size (see next section). The high-rate Ultralife cells will operate over the temperature range −40 to +70°C and are thus well suited to outdoor use in all climates. The cells incorporate safety vents which activate at low over-pressures if the cell is severely heated, for example, as a result of short-circuiting.

Ultralife Batteries, Inc. also offers a range of wafer-thin, rectangular, lithium–manganese-dioxide cells. The thickness of these cells is between 0.5 and 4.6 mm and the capacity varies from 27 to 2300 mAh (Figure 6.7). The cells have specific energies which approach 300 Wh kg^{-1}, a stable and flat discharge voltage curve, a shelf-life of up to 10 years, and an operating temperature range of −20 to +50°C. The thin cells are used as power sources for such applications as tracking systems for container freight, flexible printed-circuit boards, and smart cards.

Silver chromate, Ag_2CrO_4, is another material which has been employed extensively as a solid positive electrode in lithium primary cells. The resulting cell has an open-circuit voltage of 3.5 V and discharges according to the following reaction.

$$2Li + Ag_2CrO_4 \longrightarrow Li_2CrO_4 + 2Ag \tag{6.2}$$

Lithium–silver-chromate cells are manufactured in both button and coin

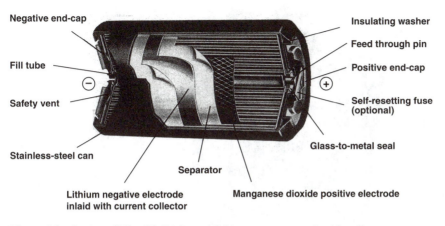

Negative end-cap

Fill tube

Safety vent

Stainless-steel can

Lithium negative electrode
inlaid with current collector

Separator

Manganese dioxide positive electrode

Insulating washer

Feed through pin

Positive end-cap

Self-resetting fuse
(optional)

Glass-to-metal seal

Figure 6.6   *Design of Ultralife (high-rate) lithium–manganese-dioxide cell*
(By courtesy of Ultralife Batteries, Inc.)

configurations, as well as in the larger prismatic form. The energy density is over 500 Wh dm^{-3} to a cut-off voltage of 2.5 V. The prismatic cells have been employed widely to power heart pacemakers on account of their excellent reliability when drawing low continuous currents, as well as their very low self-discharge rate ($<1\%$ per annum). Nowadays, the cells have largely been superseded for this application by the lithium–iodine, P2VP system (see Section 7.1, Chapter 7).

It should be noted that all lithium cells with organic electrolytes are potentially hazardous devices, with the possibility of fire or explosion if mistreated. Cells should not be heated above 100°C, short-circuited, dismantled, or operated with other battery types. Care should also be taken to insert the batteries the correct way round. Under no circumstances should attempts be made to recharge primary lithium batteries. When used as memory back-up devices with a main power source, reverse-flow protection should be ensured by inserting two silicon diodes in series between the main power source and the lithium cell. Such a circuit safeguards against recharging of the lithium cell by the main power source, a very dangerous situation. Most lithium primary cells, other than miniature cells, are best replaced by trained technicians who appreciate the potential hazards.

6.4   CELLS WITH LIQUID POSITIVE ELECTRODES

High-rate, lithium primary cells have also been developed using liquid positive electrodes. These are of two types: the 'dissolved electrode' cell and the 'inorganic liquid electrode' cell. In both designs, the positive active-material is in direct contact with the lithium-metal negative, with

Figure 6.7   *Ultralife 'Thin Cells^{TM}'*
                (By courtesy of Ultralife Batteries, Inc.)

which it reacts superficially to form a passivating layer.

Dissolved positive electrode. This consists of a solution of sulfur dioxide in an electrolyte of lithium bromide (LiBr) dissolved in acetonitrile (CH_3CN). The discharge reaction of the cell is:

$$2Li + 2SO_2(\text{diss.}) \longrightarrow Li_2S_2O_4 \quad V° = +3.0\,V \tag{6.3}$$

The partial pressure of the dissolved sulfur dioxide is $\sim 300\,kPa$ at room temperature, and rises to 3 MPa at 100°C. This necessitates a pressure housing for the cell. Alternatively, a liquid SO_2 solvate of $LiAlCl_4$, such as $LiAlCl_4·xSO_2$, may be used as the electrolyte. The positive current-collector is normally composed of a mixture of a carbon black and Teflon, which is cast into sheet form, pressed on to a supporting metal screen, and separated from the lithium-foil negative by a microporous polypropylene separator. This sandwich is spirally-wound into a nickel-plated steel can.

Lithium–sulfur-dioxide cells have gravimetric and volumetric energy densities up to 330 Wh kg^{-1} and 500 Wh dm^{-3}, respectively, *i.e.* several times the corresponding values for alkaline-manganese counterparts. Their on-load discharge curve is very flat at 2.8 to 2.9 V, and is little dependent on the current drawn because of the low internal resistance of the spirally-wound design. The temperature range over which the cells can operate is from -50 to $+70°C$, which is a much greater range than for any cell with an aqueous electrolyte. Lithium–sulfur-dioxide cells also have a long shelf-life, but may exhibit a short voltage delay on start-up as the passivating layer on the lithium negative electrode has first to be broken down. Military applications include naval sonar buoys, army field communications systems, and night vision equipment.

Inorganic liquid positive electrode. This employs thionyl chloride ($SOCl_2$) or sulfuryl chloride (SO_2Cl_2) as the active material, with dissolved $LiAlCl_4$ salt to provide the conductivity. Thionyl chloride is the more usual positive and the discharge reaction of the cell is:

$$4Li + 2SOCl_2 \longrightarrow 4LiCl + SO_2(diss.) + S \quad V° = +3.6\,V \qquad (6.4)$$

The sulfur dioxide, which is formed on discharge, is mostly soluble in the thionyl chloride and thus there is no large build-up in pressure. In addition, the soluble SO_2 may be reduced further at the positive electrode to lithium dithionite ($Li_2S_2O_4$).

The cell has an operating voltage of 2.9 V at a current density of 200 mA cm^{-2}. Like the lithium–sulfur-dioxide system, it has a high specific energy (330 Wh kg^{-1}), an even higher energy density (>700 Wh dm^{-3}), a good low-temperature performance, and a long shelf-life. Consequently, the cell is very suitable for military use in cold climates and for power applications in spacecraft.

Lithium–thionyl-chloride cells have been manufactured in all sizes, from miniature cells to units of up to 10 000 Ah. The smaller cells have found a wide range of uses which include power for printed-circuit boards, biotelemetry, meteorology, and emergency location signals. The larger batteries have been employed for torpedo propulsion and as a power source for missiles. The torpedo battery is reported to have a specific energy of 250 Wh kg^{-1} and a power density of 1900 W kg^{-1}. It is thus one of the most powerful batteries developed for use at sea.

Chapter 7

Specialized Primary Batteries

There are some applications for batteries where the requirements are so specialized that it has been necessary to develop new types of battery. Here, we consider primary batteries of this genre. Specialized secondary batteries are discussed in Chapters 11 and 12.

7.1 BATTERIES IN MEDICINE

Battery-powered devices are playing an increasingly important role in medicine. These fall into three categories:

- devices which are implanted in the body (cardiac pacemakers, defibrillators, neurostimulators);
- devices which are worn by the patient (cardiac or blood pressure monitoring devices, hearing aids, ambulatory external drug delivery systems);
- devices used by medical staff (emergency portable defibrillators, cordless surgical tools), or by patients (electric wheelchairs).

Those in the third category generally employ conventional batteries. Devices in the second category may use either conventional batteries or batteries developed primarily for this purpose (*e.g.* zinc–air cells for hearing aids; see Section 5.3.3, Chapter 5). It is in the first category of implantable devices that the need for specialized batteries arises. The specifications for such batteries include maximum possible gravimetric and volumetric energy densities, so as to give long operational life within the limited space available for the battery and so as not to make the device too heavy for the patient. Battery replacement involves further surgery which is both expensive and traumatic for the patient and, therefore, the operational life of the battery should be at least five years, preferably ten years. The battery must also be totally reliable and totally safe. Less demandingly, the current drains required are generally very small and the device operates exclusively at body temperature.

Early pacemakers employed zinc–mercury-oxide cells (see Section 5.3.2, Chapter 5) or even, in a few instances, thermoelectric generators powered by the heat released by the radioactive decay of plutonium-238. The latter performed well and had an indefinite life, but had certain other drawbacks in respect of the proliferation and dissemination of radioactive material and did not therefore come into widespread use. Zinc–mercury-oxide cells were used throughout the 1960s to power pacemakers. While this system helped to make possible the development of the pacemaker, it did have some disadvantages. The generation of hydrogen as a by-product did not allow the hermetic sealing of the cell. Moreover, there was significant self-discharge at blood temperature (37°C) and the end of discharge was abrupt with little or no forewarning. Finally, catastrophic failure often occurred through the development of an internal short-circuit which was caused by dendrite penetration of the separator. At this period, the average life of a pacemaker was only two to three years, and four out of five replacements were necessitated by failed batteries. Clearly, there was a strong incentive to develop a better, specialized battery for pacemakers.

Lithium cells offered obvious advantages on account of their high voltage and flat discharge curves, as well as high gravimetric and volumetric energy densities. Many lithium–silver-chromate cells were implanted in the 1970s (see Section 6.3, Chapter 6), as well as lithium–cupric-sulfide cells and lithium–thionyl-chloride cells. Eventually, it was realized that the ideal lithium cell for this application would be an all-solid-state device in which no liquid electrolyte was present. Already by 1970, solid electrolytes had been developed that had a high conductivity for both silver ions and sodium ions, and during the 1970s the science of solid-state electrolytes advanced rapidly.

The breakthrough for heart pacemakers came with the invention of the lithium–iodine, poly-2-vinylpyridine (P2VP) battery. The reaction between lithium and iodine was thought to be promising because the cell has an open-circuit voltage of 2.80 V at 25°C. The discharge reaction is:

$$\text{Li(s)} + \tfrac{1}{2}\text{I}_2\text{(s)} \longrightarrow \text{LiI(s)} \quad V^\circ = +2.80 \text{ V} \tag{7.1}$$

The problem with this all-solid-state reaction is that neither iodine (the positive electrode) nor lithium iodide (the solid electrolyte) has a significant electrical conductivity, and so a cell fabricated in the usual manner will provide no current. This problem was overcome when it was discovered that iodine reacts with P2VP to give a charge-transfer complex which has an electronic conductivity at least five orders of magnitude greater than that of pure iodine. The optimum composition lies at around three to five moles of iodine per mole of P2VP. The reaction product is a highly viscous, tar-like material which may be cast on to a current-collector. When this positive electrode is brought into contact with a sheet of lithium foil (negative electrode), an exceedingly thin layer of lithium iodide is formed at the interface between the two electrodes and this acts as both the electrolyte and the separator. Because the layer is very thin, the resistance is kept to a sufficiently small value so that the cell may be used as a low-current device. When the external circuit is completed, lithium ionizes at the Li | LiI interface and lithium ions diffuse through lattice defects in the thin electrolyte layer to react with iodine at the positive electrode. Thus, as discharge proceeds, the LiI electrolyte layer is gradually built up at the LiI | I_2,P2VP interface.

Cells of the above type are only capable of providing very small currents, typically 1 to 5 μA. A steady current of 2 μA for 10 years would require a discharge of 175 mAh, a perfectly realizable capacity. In practice, several times this capacity would be required (or, alternatively, the current or duration reduced) to allow for the fact that not all the iodine would be consumed before the cell resistance rose to an unacceptable level. This steady increase in cell resistance is partly due to progressive thickening of the electrolyte layer and partly to consumption of the iodine. Nowadays, most cardiac pacemakers are only required to operate 'on demand', *i.e.* the heart is stimulated only after the device detects missed heartbeats. Consequently, the capacity required for a 10-year battery life is now considerably less than when pacemakers were operated continuously.

Lithium–iodine cells were developed first as button cells of capacity 100 to 250 mAh for watches and pocket calculators. Larger cells, up to 850 mAh, were manufactured for mounting on printed-circuit boards. These larger cells have a specific energy of $\sim$ 120 Wh kg^{-1} and an energy density of $\sim$ 600 Wh dm^{-3} when discharged at low rates. It was the high reliability of these cells, coupled with the lack of liquid electrolyte and the absence of gassing, which made them attractive for use in cardiac pacemakers.

A typical pacemaker battery has a volume of 6 ml and mass of 22 g. At a discharge rate of 10 μA, the practical capacity of the cell is around 1.5 Ah. The cell is enclosed in a stainless-steel case of all-welded construction and has a glass-to-metal seal for the electrical feed-through. It is assembled in a dedicated dry-room of low humidity. Three generations of cardiac pacemaker are shown in Figure 7.1, together with the improvements

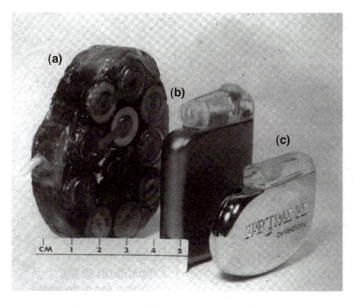

Figure 7.1 *Development of cardiac pacemakers. (a) Mid-1960s model with 10 zinc–mercury-oxide cells encased in epoxy resin; (b) 1980 model; (c) 1988 model encased in stainless-steel and powered by a lithium–iodine, poly-2-vinylpyridine battery* (By courtesy of Batteries International)

which have been made in design and in volume reduction. An X-ray of a patient's chest with the pacemaker *in situ*, implanted under the shoulder blade, is presented in Figure 7.2. About twenty companies, world-wide, make pacemakers and around half a million of the devices are implanted annually.

Although solid-state lithium–iodine cells are a great improvement over zinc–mercury-oxide cells for use in pacemakers, they still have some limitations. Their volumetric energy density is excellent, but their gravimetric energy density is not so good, since iodine is rather dense and the cell is contained in a stainless-steel case. Also, the cell resistance rises dramatically during discharge due to build-up of the LiI discharge product. For these reasons, consideration is being given to reverting to liquid-electrolyte lithium cells with carbon monofluoride positive electrodes. By containing such cells in a titanium enclosure, it should be possible to double the gravimetric energy density, while retaining the favourable volumetric energy density. In addition, the cell resistance should be steady at $< 50\,\Omega$ throughout discharge instead of rising to thousands of ohms as with lithium–iodine cells. On the other hand, thorough testing will be necessary to ensure that there are no other problems associated with these cells.

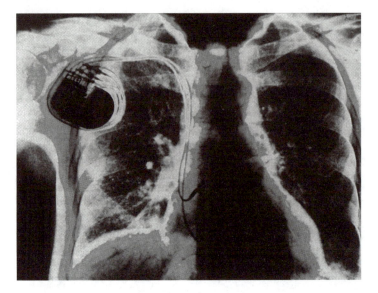

Figure 7.2   *X-ray of patient's chest with pacemaker implanted*
(By courtesy of Batteries International)

Another medical application of batteries is in defibrillators. Ventricle
fibrillation occurs when the heart rate becomes so fast that it stops
pumping and quivers uncontrollably. A severe electrical shock may jolt
the heart back to a normal rhythm, and most hospital casualty depart-
ments, coronary units and ambulances have defibrillators which apply
the shock through external electrodes. As this treatment causes severe
trauma to the patient, implantable defibrillators were introduced in the
1980s. The implanted batteries are required to supply pulses of ~ 25 J
directly to the heart. Such power pulses are too high for solid-state
batteries, and so batteries with organic liquid electrolytes have been
developed for this application. Today, most implanted defibrillators use
lithium batteries with silver vanadium oxide (AgV_2O_5) positive electrodes.

There are several other implantable electrical devices, either in use or
under development. Bone stimulators encourage the knitting of broken
bones in cases of non-union. Neurostimulators may relieve chronic back-
ache or reduce the severity of epileptic seizures. Left-ventricular assist
devices provide a stand-by treatment for patients awaiting heart trans-
plants. Drug delivery systems provide carefully controlled administration
of medicines inside the patient. Altogether, there are high hopes for the
future of electrical devices in the medical field and many of these will
require specialized batteries for their operation.

7.2 SEA-WATER ACTIVATED BATTERIES

There are many applications for batteries at sea. Among these may be cited:

- rescue lights on life jackets and life-rafts;
- navigation lights on buoys;
- radio locator beacons;
- sonobuoys;
- meteorological systems;
- divers' searchlights;
- markers for submarine escape units and emergency lighting;
- torpedo propulsion;
- underwater defence systems.

Conventional batteries may be used for many of these applications, but often it is preferable to employ batteries which are activated by sea-water.

Sea-water, provided it is of adequate salinity, makes a good battery electrolyte. One of its principal attractions is that it is only introduced into the battery at the point of use and, before that, the battery is stored dry. This eliminates most causes of self-discharge and gives the battery a long or indefinite shelf-life. For this reason, such batteries are often referred to as 'reserve batteries'. Long shelf-life is particularly important for emergency equipment which, by definition, is used only occasionally. Similarly, military equipment such as torpedoes require an indefinite shelf-life while being available for use at very short notice.

Magnesium is invariably used as the negative electrode in sea-water batteries. This has a much higher specific capacity (2.20 Ah g^{-1}) than zinc (0.82 Ah g^{-1}). The magnesium is usually in the form of an alloy with small amounts of aluminium (3 to 6 wt.%) and zinc (1 wt.%). The alloys are electrochemically active and can be rolled readily.

There is a choice of possible positive electrodes to use in conjunction with magnesium negatives, *viz.* silver chloride (AgCl), lead chloride ($PbCl_2$), cuprous chloride (Cu_2Cl_2), or lead dioxide (PbO_2). Each has its advantages and disadvantages, although silver chloride is often the preferred electrode system.

Silver chloride. This is an unusual inorganic material in that it can be cast and rolled like a metal, an attribute which makes for ease of fabrication into sheets as required for a battery. The magnesium–silver-chloride battery does not normally incorporate a conventional type of separator, but has an open configuration to allow the sea-water to flow through it. The positive and negative electrodes are held apart by small plastic spacers which have minimal impact on the flow of water.

The magnesium–silver-chloride cell may be represented as Mg(s) | sea-water | AgCl(s),C(s). The discharge reaction is:

$$Mg(s) + 2AgCl(s) \longrightarrow Mg^{2+}(aq.) + 2Cl^-(aq.) + 2Ag(s) \qquad (7.2)$$

The standard voltage for this cell is 2.59 V, but the open-circuit voltage in sea-water is only 1.7 to 1.9 V and the on-load voltage is 1.5 V. This departure from the theoretical voltage is thought to result from the build-up of an insoluble surface layer on the magnesium and this gives rise to the generation of considerable heat energy. The discharge curve is flat and not greatly dependent on current density. There is little polarization since the reaction product, magnesium chloride, is soluble in sea-water. A side-reaction is the chemical corrosion of magnesium by sea-water. This leads to a sludge of magnesium hydroxide and magnesium oxychloride that may, to some degree, impede the discharge reaction. Hydrogen is evolved in this corrosion reaction. Many advantages are claimed for the magnesium–silver-chloride cell, namely:

- ease of fabrication;
- indefinite storage life at -50 to $+70°C$;
- tolerant to storage under conditions of high humidity;
- wide operating conditions (0 to $+35°C$, in salinity of 1.5 to 3.6 wt.%);
- high specific energy, *i.e.* up to 165 Wh kg^{-1};
- variable duty cycle (from seconds to days);
- rapid activation (<1 s);
- wide current capabilities, *i.e.* pulsed or continuous;
- non-hazardous in storage or use;
- resistant to thermal and mechanical shock;
- the load voltage (1.5 V) allows the use of commercial lamp bulbs.

Cells and batteries may be custom-designed to suit the application. Electrodes may be either flat or spirally-wound. The downside of these batteries is the rather high cost of silver chloride compared with the other negative electrode materials.

Lead chloride. This compound is considerably cheaper than silver chloride, but unfortunately cannot be cast and rolled. The electrode has therefore to be prepared by compressing lead chloride powder on to a metal gauze, or by an alternative conventional process. The discharge reaction of the cell is:

$$Mg(s) + PbCl_2(s) \longrightarrow Mg^{2+}(aq.) + 2Cl^-(aq.) + Pb(s) \qquad (7.3)$$

The cells are of similar design to those with silver chloride electrodes, but do possess certain disadvantages. For example, the on-load voltage is only 1.08 V compared with 1.5 V for the silver chloride system, the lead chloride electrodes do not perform well at high current densities, and the cells may only be used for low-drain applications.

Cuprous Chloride. The magnesium–cuprous-chloride cell has much the same design and properties as the magnesium–lead-chloride counterpart, but the discharge voltage is higher (1.3 to 1.4 V). Again, the cell is limited to low discharge rates. A further complication is that the cell must be hermetically sealed for storage as cuprous chloride is hygroscopic. To put the cell into use, it is first necessary to break the seal so that sea-water may enter.

Lead dioxide. The magnesium–lead-dioxide cell has an operating voltage of 1.5 to 1.6 V. The cell reaction may be represented simply as:

$$Mg(s) + PbO_2(s) + 2H_2O(l) \longrightarrow Mg(OH)_2(s) + Pb(OH)_2(s) \qquad (7.4)$$

The positive electrodes are conventional pasted plates of lead dioxide, as used in lead–acid batteries (see Section 8.1, Chapter 8). The electrodes operate less efficiently than their metal halide counterparts when immersed in a sea-water electrolyte. To reduce the cell resistance, a thin paper separator is used in place of a simple spacer. The separator tends to retain the electrolyte, rather than allow it to flow through the battery. The cell resistance falls as the temperature rises. This compensates for the voltage drop as discharge proceeds and the resulting discharge curve is essentially flat. As with lead chloride cells, the current density is limited to low values, *viz.* ~ 15 mA cm^{-2}. There is, however, an advantage with the lead dioxide design. If the porous paper separator is pre-saturated with brine and allowed to dry before assembling the cell, it is possible to produce a unit which can be activated by fresh water. This feature could be useful, for instance, on the Great Lakes in Canada.

A generic problem with sea-water batteries which do not employ a separator is that when cells are connected in series, there is considerable current leakage ('shunt currents') from cell to cell on account of the open configuration. For this reason, batteries intended for low-rate discharge (*i.e.* long operational life) should be connected in parallel. This limits the output voltage to that of a single cell. If this is not acceptable, it is possible to utilize a d.c.–d.c. converter to boost the voltage.

An interesting concept, which is being developed in Norway, is that of using oxygen dissolved in the sea as the oxidant in conjunction with a fuel-cell type of positive electrode. This necessitates a cell with a very open configuration and a good throughput of sea-water, so that the supply of oxygen shall be adequate. Very large cells have been built for powering unattended marine lights over long periods. The magnesium-alloy negative electrodes are rods of 1.0 m in length, 190 mm in diameter, and 50 kg in weight. Around these are rigid positive electrodes, in the form of concentric coils, made of either copper or carbon fibre. The main advantage of carbon fibre over copper in sea-water cells is the higher voltage obtained, *i.e.* 1.4 to 1.6 V *vs.* 0.8 to 1.0 V. The disadvantage is that the carbon-fibre electrodes are subject to more severe fouling by algae and other marine life. For this reason, batteries with carbon negatives are used in deep-water applications where fouling is less of a problem. Such a battery has been employed to propel an unmanned submersible.

Batteries with copper negatives are preferred for surface applications such as marine lights. Even so, inorganic fouling occurs through the formation of deposits of calcium carbonate and insoluble copper salts. The complete power supply for a marine light consists of the sea-water cell, a d.c.–d.c. converter, and a 12 V, valve-regulated lead–acid battery (see Section 8.7, Chapter 8) which acts as a buffer store. The sea-water cell charges the lead–acid battery continuously at a low rate and the latter provides pulses of power for a flashing light during the hours of darkness. A typical sea-water cell of this type has a useable capacity of 50 000 Ah, a voltage of 0.95 V and a converter efficiency of 75%, which gives a system energy content of 35 kWh. With a cell of this size, prototype marine lights have operated unattended for two to three years before breakdown, and a target life of five years seems obtainable. At the end of this time, the cell will be cleaned of fouling and a new magnesium electrode fitted. The unit can then be seen as a 'mechanically rechargeable' battery.

7.3 TORPEDO BATTERIES

The magnesium–air battery described in the section above is of very large capacity but low continuous power output (~ 1 W) and will therefore last for several years. A torpedo propulsion battery, by contrast, has to produce a very high power output for just a few minutes. Early torpedoes were propelled by compressed air, but this limited both the range and the speed of the weapon. There were other disadvantages too, namely: the wake of gas bubbles was easily spotted, and this allowed avoidance action to be taken; the cooling which occurred when the gas expanded constituted a wastage of power.

Electrically-propelled torpedoes were introduced during World War II. One of the problems encountered was that the batteries (lead–acid types) lost charge during storage and required frequent recharging. Dry storage was not a practical option because of the time taken to introduce the sulfuric acid before firing. In a war situation, a torpedo is a 'one-shot' device and requires only a primary battery to propel it. Most torpedo firings, however, are practice shots with no explosive charge and the weapon is recovered for re-use. Rechargeable batteries are then needed and conventional lead–acid or nickel–cadmium batteries are a suitable choice, though with limited power and range. Rechargeable zinc–silver-oxide batteries (see Section 9.7, Chapter 9) are also used for this purpose because they are more powerful and can provide the rather few charge–discharge cycles needed for practice shots, but at considerably higher cost.

Torpedo batteries are discharged at the 5- to 10-min rate ($12C$ to $6C$ rate) and, therefore, must be very powerful. As ships become faster, it is necessary to have ever more powerful propulsion units for torpedoes. The power required is approximately proportional to the cube of the speed through the water. Thus, for a 10% increase in speed, it is necessary to have a 33% increase in power output from the battery. The space available to accommodate the battery is, however, strictly limited. Both the diameter and the length of the weapon are essentially fixed and space has to be made for both the high explosive charge and the guidance and control system. It follows that the power density of the battery (W dm^{-3}) is a key parameter, along with reliability, safety in operation, long shelf-life, resistance to shock and vibration, and rapid activation. High energy density (Wh dm^{-3}) is also important to give the desired range. For batteries generally, high power and high energy tend to be mutually exclusive. Consequently, specifications for the power units of torpedoes are most demanding. Since the 1970s, warshot torpedoes have been propelled principally by zinc–silver-oxide or magnesium–silver-chloride reserve batteries, although there has also been some use of lithium–thionyl-chloride batteries (see Section 6.4, Chapter 6).

7.3.1   Zinc–Silver-oxide Reserve Battery

This is a reserve battery, in that it is stored in the dry state and is activated by injecting the electrolyte (36 wt.% potassium hydroxide solution) into the cells at the point of use. The electrolyte is stored separately in an external reservoir and activation is brought about by rupturing a diaphragm and causing the solution to flow into the cell under gravity or under pressure. To ensure a high-rate discharge, the plates are thin, typically 0.7 mm for the negative and 0.55 mm for the positive. Cells can have between 15 and 45 pairs of plates per cell. A triple separator is used, with each component having a defined role, and has an overall thickness of 0.4 mm. The negative plate consists of spongy zinc on a mesh of silver-coated copper, while the positive plate is porous, sintered silver deposited on a pure silver mesh. The positive is formed electrochemically all the way through to the higher oxide Ag_2O_2.

The discharge reaction of the cell takes place in two stages, namely:

$$Ag_2O_2 + Zn + H_2O \longrightarrow Ag_2O + Zn(OH)_2 \quad V° = +1.85\,V \qquad (7.5)$$

$$Ag_2O + Zn + H_2O \longrightarrow 2Ag + Zn(OH)_2 \quad V° = +1.60\,V \qquad (7.6)$$

In practice, at the high discharge rates employed, the two steps in the discharge curve are smeared out and most of the energy is extracted at around 1.4 V. A discharge curve for a 45-plate cell (weight: 4.5 kg) at the 16-min rate is shown in Figure 7.3. The discharge current is 1210 A, which corresponds to a current density of 125 mA cm^{-2}. Even higher discharge rates are possible. The flat discharge curve ensures that the torpedo maintains its speed throughout its travel.

One problem with zinc–silver-oxide reserve batteries is that they are unstable following activation and, if not discharged, undergo self-discharge through internal short-circuiting after a delay period which depends on the temperature. The batteries are used particularly in heavy-weight torpedoes, as fired from ships or submarines.

7.3.2   Magnesium–Silver-chloride Reserve Battery

The essential features of the magnesium–silver-chloride battery have been described above in Section 7.2. For torpedo batteries, the design

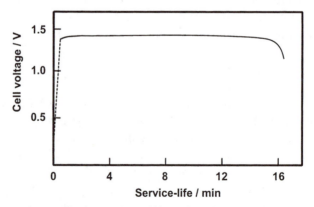

Figure 7.3   *Discharge of high-rate zinc–silver-oxide cell at 20°C*
(By courtesy of Research Studies Press Ltd.)

must maximize the power density. This is achieved by using a highly-reactive magnesium alloy and thin electrodes (0.3 mm). The bipolar design of the battery is shown in Figure 7.4. The positive plate (silver chloride) has a series of small glass beads attached to it that serve as spacers to separate the plate from the adjoining negative (magnesium alloy). A thin (0.02 mm) sheet of silver foil is placed between one cell and the next. This acts as an inter-cell connector (bipole), and also as a mechanical barrier to prevent sea-water from adjacent cells mixing and giving rise to unwanted shunt currents. For the same reason, the circumference of the plates is embedded in a resin case. Water flows through the spaces between the electrode pairs at a steady rate so as to remove heat, magnesium hydroxide sludge and the hydrogen liberated by corrosion. There are between 100 and 120 series-connected cells in a battery. The energy density of the battery stack is ~ 300 Wh dm^{-3} at the 6-min rate, which corresponds to a power density of 3 kW dm^{-3}.

At the very high discharge rates required by torpedos, the battery voltage does decline as discharge proceeds. On the other hand, the voltage is temperature sensitive and increases with increasing temperature. By recirculating a fraction of the warm exit sea-water (containing zinc chloride discharge product) to the inlet manifold, it is possible to increase the salinity of the electrolyte (and therefore its conductivity) and also its temperature. This provides a means to compensate for the falling voltage, and thus a much flatter discharge curve is maintained, as shown in Figure 7.5. The batteries may be operated at current densities as high

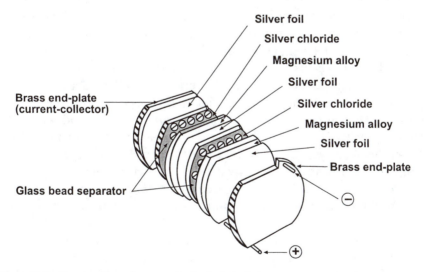

Figure 7.4 *Construction of a sea-water torpedo battery*
(By courtesy of Research Studies Press Ltd.)

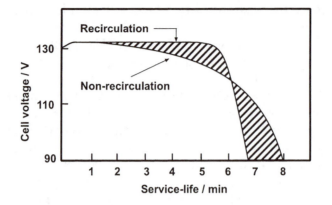

Figure 7.5 *Discharge of a sea-water torpedo battery with and without recirculation*
(By courtesy of Research Studies Press Ltd.)

as 420 mA cm^{-2}. Batteries with several hundred kilowatts of power have
been built for this application. Sea-water reserve batteries are also well
suited to lightweight torpedoes which are dropped from aircraft.

7.4 THERMAL BATTERIES

Thermal batteries are reserve batteries which are activated by the application of heat. Generally, they contain a metallic salt electrolyte which is
non-conducting when solid at ambient temperature, but which is an
excellent ionic conductor when molten. Such batteries may be designed
to have very high power outputs for short periods of time (seconds to
minutes), and to have indefinite shelf-life in storage. These are characteristics which make them ideal as power sources for weapons systems,
particularly guided missiles. The batteries are activated by the rapid
ignition of a pyrotechnic charge which is incorporated in the battery
housing. Thereafter, the battery remains active for a period which depends upon its size, thermal insulation, the ambient temperature, the
electrochemical system, and the ratio of pyrotechnic charge to battery
components. Thermal batteries were first devised in Germany towards
the end of World War II. Since then, considerable development has taken
place and they are now widely employed by the military in many countries.

The electrolyte which is used in thermal batteries is often a eutectic
mixture of lithium and potassium chlorides (LiCl : KCl) that melts
at 352 °C. Alternatively, a ternary eutectic mixture of LiCl : LiBr : LiI or
LiF : LiCl : LiBr may be chosen. Thermal batteries are normally operated

at between 400 and 540°C. At these temperatures, the salts are fully ionized and have a very high ionic conductivity while being electronic insulators. Usually, an inert solid 'filler' (*e.g.* silica or magnesia) is mixed with the electrolyte so that when the salt melts, it is retained in position as a semi-solid paste.

The electrochemical couple originally used consisted of a negative electrode of calcium metal and a soluble positive electrode of calcium dichromate or chromate dissolved in the electrolyte. For a long while, the preferred cell was Ca | LiCl : KCl | CaCrO$_4$. Nevertheless, various problems are encountered with this cell chemistry and more recently lithium negative electrodes have been employed (either pure or alloyed with aluminium, silicon, or boron) in conjunction with an insoluble positive electrode of iron disulfide. The detailed chemistry of the lithium–iron-disulfide discharge reaction is complex and involves several stages, but overall it may be represented as:

$$4Li + FeS_2 \longrightarrow 2Li_2S + Fe \qquad (7.7)$$

with a cell voltage which falls from 2.1 to 1.6 V. The lithium–iron-disulfide cells have largely superseded calcium types by virtue of having the following advantages: greater simplicity in construction; higher current capabilities, *i.e.* can be used for duties requiring long discharge lives (40- to 60-min rate) and for pulse applications (0.5-min rate); less susceptibility to efficiency-reducing and exothermic side-reactions; and greater stability and reliability under severe dynamic conditions (launch shock, vibration, acceleration, axial spin, *etc.*).

The construction of a typical thermal battery is shown in Figure 7.6. The essential design feature is that the thermal activation process shall be very rapid. To this end, it is necessary to incorporate the pyrotechnic charge in the body of the cell stack. As shown, the cell stack is built up of the sequence: metal cup | negative | electrolyte | positive | pyrotechnic charge. The metal cup acts as a bipole to join adjacent cells in series electrically. Current is collected from the electrodes at each end of the cell stack. The battery voltage depends on the number of cells in the stack, and typically ranges from 20 to 60 V.

The pyrotechnic charge normally consists of a mixture of finely-divided iron powder and potassium perchlorate (KClO$_4$) compressed into a dense pellet. The thickness of the pellet is determined by the amount of heat required to sustain the battery operation (which may last from a few tens of seconds to 20 minutes or more). The pellet is ignited electrically from an auxiliary battery by passing a current through the ignition strip which contacts each pyrotechnic charge. The time for

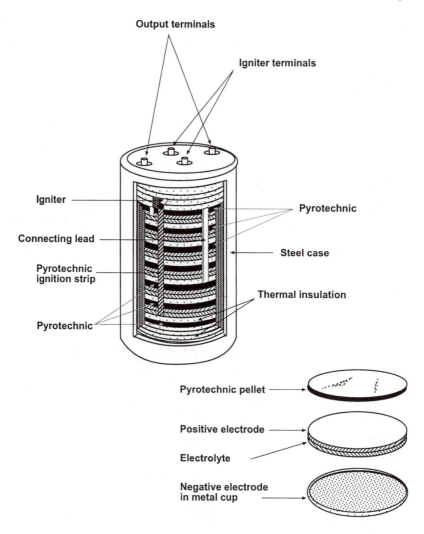

Figure 7.6 *Construction of a thermal battery*
(By courtesy of Batteries International)

activation (that is, until the electrolyte melts and a voltage appears) varies from 50 ms to several seconds according to the size and construction of the battery. As always with batteries, the design has to be optimized to meet the mission specification. Thin electrodes activate rapidly and give high power output but, having low capacity, deliver power for a shorter time. Thick electrodes support a longer duration mission, but take longer to activate and also produce less power.

Part 3

Rechargeable Batteries

Until quite recent times, there were effectively only two types of rechargeable battery available. The first, and overwhelmingly the most used, was that based upon the lead–lead-dioxide (Pb–PbO$_2$) electrochemical couple with sulfuric acid as the electrolyte – the so-called 'lead–acid battery'. The second type employed a metal negative electrode of cadmium or iron, in conjunction with a nickel oxyhydroxide ('nickel oxide') positive electrode and an electrolyte of potassium hydroxide (alkaline) solution. The cadmium–nickel-oxyhydroxide (Cd–NiOOH) and the iron–nickel-oxyhydroxide (Fe–NiOOH) batteries became known popularly and commercially as 'nickel–cadmium' and 'nickel–iron' batteries, respectively.

In the 1920s and 1930s, zinc–nickel-oxyhydroxide (Zn–NiOOH) batteries ('nickel–zinc') were also commercially available, but these had short cycle-lives and failed to gain market acceptance. Recent research and development is leading, however, to a re-emergence of this battery as a commercial entity. Similar alkaline batteries with positive electrodes of silver oxide ('silver–zinc' batteries) have also been manufactured, but their high cost and very limited cycle-life has restricted their use to specialized applications. It is only in the last couple of decades that new types of rechargeable battery, namely, the nickel–metal-hydride battery and the lithium-ion battery, have become commercially important.

A true secondary battery is expected to undergo at least 500 charge–discharge cycles, and often many more. For instance, an electric-vehicle traction battery should have a life of 1000 cycles to be commercially viable, while a battery in a low earth orbit satellite is expected to last for more than 20 000 cycles. In chemical terms, this translates into electrode reactions which are fully reversible for the desired number of cycles. The difficulty is that most batteries possess solid electrode materials and liquid electrolytes and that the electrode reactions involve solid-state

97

diffusion which results in phase changes and recrystallization. A typical positive electrode reaction would be:

$$\text{Solid A} + e^- \underset{\text{Charge}}{\overset{\text{Discharge}}{\rightleftarrows}} \text{Solid B} + \text{Anion}$$

For example:

$$NiOOH + H_2O + e^- \underset{\text{Charge}}{\overset{\text{Discharge}}{\rightleftarrows}} Ni(OH)_2 + OH^-$$

From the viewpoint of the solid-state chemist, the requirement to reverse this reaction *quantitatively* for hundreds or thousands of times is an exceedingly demanding schedule. The severity of the specification is apparent when one considers the many possible side-reactions which lead to battery deterioration and failure. These include:

- densification of active material with loss of porosity;
- expansion and shedding of active material from electrode plates;
- progressive formation of inactive phases, which electrically isolate regions of the active material;
- growth of metallic needles at the negative electrode, which give rise to internal short-circuits;
- gassing of electrode plates on overcharge, which cause disruptive effects;
- corrosion of current-collectors, which results in increased internal resistance;
- separator dry-out through overheating.

Such degradation processes may result in precipitous battery failure, through an internal short-circuit, or in progressive loss of capacity and performance. Generally, the degeneration steps are interactive and accumulative, so that when the performance starts to deteriorate, it soon accelerates and the battery becomes unusable. Despite this gloomy prognosis, some remarkable successes have been achieved in designing batteries with a long cycle-life ($\sim$ 1000 cycles) for several different chemistries. One particular battery – the nickel–hydrogen battery – has been demonstrated to last for over 20 000 cycles and has been the preferred battery for use in satellites.

Research into rechargeable batteries is an active field. Chapters 8 to 10 describe, respectively, advances in lead–acid batteries, alkaline electrolyte batteries, and rechargeable lithium batteries. Chapter 11 summarizes briefly several different types of novel battery which are either at an advanced stage of development or are commercially available in small numbers from pilot plants. Finally, Chapter 12 provides a more detailed discussion of four particular applications for secondary batteries to illustrate how technical and operational factors both contribute to defining a battery specification. It is seen that the specification may vary widely for applications which, superficially, are quite similar.

Chapter 8

Lead–Acid Batteries

8.1 HISTORY OF THE LEAD–ACID BATTERY

The lead–acid battery was invented in 1859 by the French electrochemist Gaston Planté who made a spiral roll of two sheets of pure lead, separated by a linen cloth, and immersed the assembly in a glass jar which contained sulfuric acid solution (see Figure 1.2, Chapter 1). On passing a current between the plates, and reversing it periodically, the surface of the positive plate was gradually converted into lead dioxide (PbO_2) and that of the negative plate to spongy metallic lead. After a while, the conversion was such that the charged cell was capable of sustaining a useful discharge current. Improved cells of the Planté type, but using flat as opposed to spirally-wound plates, are still manufactured and used today for specific applications. In particular, they are preferred in the field of stand-by power for those situations where the duty cycle is occasional and light, but where long life and reliability are prime considerations.

The next major step in the development of the lead–acid battery was made by Fauré who, in 1881, coated the lead sheets with a paste of lead oxides, sulfuric acid and water. On 'curing' the plates at a warm temperature in a humid atmosphere, the paste changed to a mixture of basic lead sulfates which adhered to the lead electrode. During charging, the cured paste was converted into electrochemically active material (or the 'active mass') and thereby gave a substantial increase in capacity compared with the Planté cell. Again, the positive active-mass in the charged state is lead dioxide and the negative active-mass is metallic lead. Soon the idea developed of cutting rectangular holes out of the lead plates to lighten their weight and also to provide receptacles into which the paste could be packed. So was born the modern 'pasted-plate battery' as employed, for instance, in motor vehicles (the 'automotive battery'). This is by far the most common type of lead–acid battery in use today. The most recent development in pasted-plate batteries is that of variants where the need for top-up with distilled water is much reduced ('low-maintenance' types) or is eliminated altogether ('maintenance-free' types). Such technology is discussed in Sections 8.3.1 and 8.7.

One of the drawbacks of the pasted-plate battery is that its cycle-life is reduced considerably when it is subjected to regular deep discharge, *e.g.* when powering an electric vehicle. In part, this is attributable to expansion and contraction of the active mass, a process which causes the paste to become progressively dislodged from the plates and fall to the bottom of the cell. Loss of plate material reduces the capacity of the cell and, ultimately, may give rise to an internal short-circuit between the positive and negative plates. This paste-shedding problem, which is particularly acute for the positive plate, can be mitigated by wrapping each plate in a fibre-glass retainer and by compressing the total cell-assembly. An alternative strategy is to pack the positive active-material into porous tubes which each contain a central spine of lead alloy as the current-collector, *i.e.* the so-called 'tubular-plate battery' (see Section 8.6).

Conventional lead–acid batteries are 'flooded'. That is, the electrolyte is a free liquid to a level above the top of the plates and above the busbars. This has the disadvantage that the cells have to be vented to release the gases liberated during charging, namely, oxygen at the positive electrode and hydrogen at the negative. The consequence of this venting is that the batteries may be used only in the upright position, otherwise leakage of sulfuric acid takes place. Also, the released gases carry a very fine mist of sulfuric acid which is highly corrosive. For many years, scientists attempted to develop 'sealed' batteries which would be safe under all conditions of use and abuse. At first, such attempts revolved around the catalytic recombination of the gases within the battery; this approach has not been notably successful (see Section 8.7). Success came, however, with the invention of the valve-regulated lead–acid (VRLA) battery. In this design, oxygen is reduced ('recombined') back to water, at the negative plate. A corresponding recombination cycle for hydrogen does not exist because oxidation of the gas at the positive electrode is too slow. This, together with the fact that oxygen recombination is not complete (the efficiency is typically 95 to 99%), requires each cell to be fitted with a one-way valve as a safeguard against excessive pressure build-up – hence, the term 'valve-regulated'. The VRLA battery can be employed in any orientation, and thus gives equipment design engineers a much greater degree of flexibility.

In parallel with these advances, there has been a continuing scientific effort to understand the exact nature of the processes taking place in the lead–acid battery and, particularly, the factors which limit its cycle-life and result in ultimate failure. This is a complex issue and is bound up with the design of the battery, its materials of construction, the quality of the build, and the conditions of use. In parallel with these scientific studies, there have been steady improvements – both in the materials employed and in the design of the batteries. All these developments, which are still on-going even after more than a century, have led to new versions of the battery with improved performance and, in real terms, reduced manufacturing costs.

The history of the applications for lead–acid batteries is also interesting. The first major market was for stand-by batteries to provide emergency power to essential equipment in electricity-generating stations and at other critical sites, such as hospitals. Since then, the requirements for such batteries have expanded hugely, as described in Section 1.2, Chapter 1. Towards the end of the 19th century, electric cars appeared on the roads. Before World War I, battery electric submarines were developed. Batteries also began to be used for illumination in railway coaches, as well as for powering railway signalling systems, the electrical equipment of ships, and radio receiving–transmitting equipment. With the advent of the internal combustion engine, the lead–acid battery was first employed in road vehicles for lighting, then later also for engine starting, and now additionally for the whole range of electrical duties expected in the modern vehicle. The market for traction batteries has expanded, particularly into off-road vehicles, as discussed in Section 1.2, Chapter 1. In almost all cases, it is the lead–acid battery which predominates when the requirement is for stored energy of more than a few hundred watt-hours. For these larger battery applications, no other battery is yet able to compete on cost grounds, although batteries based on other chemistries are rapidly catching up, as is described in later chapters.

8.2 MODE OF OPERATION

As mentioned in Section 2.1, Chapter 2, the lead–acid battery is unusual among rechargeable batteries in that the electrolyte, *viz.* sulfuric acid (H_2SO_4), takes part in the electrode reactions and is not merely a conductor of ions. The discharge and charge reactions are given by Equations (2.3) to (2.5) in Chapter 2; lead sulfate ($PbSO_4$) is formed as the discharge product at both plate polarities. The sulfuric acid is consumed during discharge and regenerated during charge. This allows the state-of-charge to be determined by measuring the relative density of the electrolyte. When fully charged at 25 °C, the relative density is ~ 1.28, which corresponds to ~ 38 wt.% H_2SO_4, and falls to a value of ~ 1.1 when discharged (~ 16 wt.% H_2SO_4), see Figure 8.1. The specific conductance of

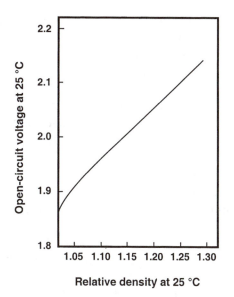

Figure 8.1 *Dependence of open-circuit voltage of lead–acid cell on relative density of electrolyte at 25°C*

the acid varies significantly with both concentration and temperature. At 20°C, the specific conductance is 0.7 and 0.6 S cm^{-1} for 40 wt.% (charged) and 16 wt.% (discharged) acid, respectively. At 40°C, the specific conductance exhibits a maximum of 1.0 S cm^{-1} at 30 to 35 wt.% acid.

A consequence of the electrode reactions is that the pores of the positive electrode tend to be denuded of acid during discharge, to be replaced with water, and then re-filled with acid during charge. A further effect is 'acid stratification', whereby the dense, concentrated acid tends to accumulate at the base of the cell with weaker acid further up the plates. Equilibrium is restored slowly by diffusional processes. A related and well-known phenomenon is that if a car battery has been drained 'flat' while trying to start a reluctant engine, it often pays to pause for 15 min or so, and then try again. During this time, acid diffuses into the positive electrode and a powerful current pulse is available once more.

The temperature coefficient of the open-circuit voltage is small, in the range + 0.1 to + 0.3 mV per °C, and has little impact on battery performance. Rather, it is the influence of temperature on the electrolyte which is the important factor. At low temperatures, the diffusion of acid into the active mass becomes slower, and also the increased viscosity of the electrolyte slows down the convective flow of the acid. Both effects result in sharply reduced power output at sub-zero temperatures. It is well known that vehicles with weak batteries are less liable to start in cold weather. This is due to increased viscosity of the oil, which necessitates a higher current to turn over the engine, and to the slow diffusional processes which limit the time for which the high current is available. The specification for an automotive battery includes a stipulation for cold-start performance ('cold-cranking amps', CCA). Various international standards exist for defining this. The common rating gives the current, in amperes, which can be drawn from the battery continuously for a stated period at $-18°C$ before the cell voltage falls below a specified value (Table 8.1). Typical CCA values are 200 to 500 A for cars, and 700 to 1000 A for commercial vehicles. Diesel engines require more power than petrol engines of similar size. To provide such currents without heavy polarization, and consequent loss of cell voltage, it is imperative that the battery has a very low internal resistance. This is a factor which battery designers have to keep keenly in mind. To provide acceptable performance, the internal resistance of automotive batteries must be restricted to the order of milliohms.

The popularity of the lead–acid battery stems, in part, from its comparatively high open-circuit voltage (~ 2.1 V). Other positive factors are low cost, versatility and excellent reversibility of the electrochemical couple. The major limitations of the lead–acid battery lie in its excessive weight, which is a consequence of the high atomic mass of lead (207.19), and its relatively poor low-temperature performance. The theoretical specific energy of the lead–acid battery, counting reactants only, is 170 Wh kg^{-1}. In practice, however, most real batteries deliver just 30 to 40 Wh kg^{-1}. This discrepancy is accounted for by the mass of the inactive components (cell container, grids, busbars, posts, cell-connectors, terminals, separators, *etc.*), and by the poor utilization of the active materials. The latter is due to the slow acid-diffusion processes in the porous reactants, and to the low electrical conductivity of the discharge product (lead sulfate) which coats individual particles and tends to isolate them electrically. The lead sulfate also blocks pores in the active mass and thereby inhibits the supply of acid. Much of the research which has been conducted on the lead–acid battery in recent years has been directed towards understanding and improving the poor utilization of the active mass.

Table 8.1 *Standard conditions for measuring cold-cranking current at* $-18°C$

Standard	Duration (s)	End voltage (V per cell)
SAE J537	30	1.2
IEC	60	1.4
DIN 435–39	30	1.5

8.3 AUTOMOTIVE BATTERIES

Most automotive batteries are composed of six series-connected cells in a monobloc, to give a nominal output of 12 V. Small motorcycles sometimes use 6 V (3 cell) batteries. Overwhelmingly, automotive batteries have flooded-electrolyte cells with Fauré pasted plates; only a few VRLA designs are offered. By contrast, VRLA motorcycle batteries are quite common. Fundamentally, the modern automotive battery is no different from that of 100 years ago. There are, however, profound differences in terms of the materials of construction, design, and performance. Some of the improvements are of an engineering rather than a scientific nature, and make use of new materials as they became available. For instance, early glass containers were replaced progressively by a bituminous composition, then by moulded rubber, and then by polypropylene. The use of polypropylene not only reduced the weight, but also allowed the development of heat-sealed, case-to-cover assemblies, and of cell interconnects which pass internally through the cell wall rather than externally across the top of the battery. The through-the-wall interconnects gave a further weight saving. The design of a 12 V monobloc from the 1930s is compared, in Figure 8.2, with one from the 1990s. Improvements, mostly of an engineering and design nature, have reduced the mass of a 40 Ah automotive battery over this period from 21 to 11 kg. The future trend will be towards VRLA batteries.

8.3.1 Positive-plate Alloys

Early grids were based on lead–antimony alloys which contained up to 8 wt.% Sb. The role of the antimony is to strengthen and harden the lead through the formation of a finely-dispersed eutectic phase at the grain boundaries. The tensile strength, yield strength and creep resistance of the alloy all increase with antimony content. In addition, the alloying agent improves both the castability of grids and the adherence of the paste to the plate. Unfortunately, the addition of antimony has attendant disadvantages. During battery charge, the antimony dissolves (corrodes) progressively from the positive grid, diffuses through the electrolyte, and deposits on the negative electrode where it lowers the overpotential for hydrogen evolution. This results in greater gassing rates, excessive loss of water and the concomitant need for more frequent maintenance, as well as increased self-discharge. To overcome this problem, a range of low-antimony alloys was developed. Initial attempts to use such alloys led to cracking of the grids on casting, but this was circumvented by the addition of grain-refining agents such as arsenic, copper, selenium, sulfur,

(a)

(b)

Smart, easy to clean, polypropylene
container and lid heat-sealed
together to provide a join equal
in strength to the material itself.

Maintenance-free, but vent plugs
can be removed for topping-up
in event of electrical malfunction
on vehicle.

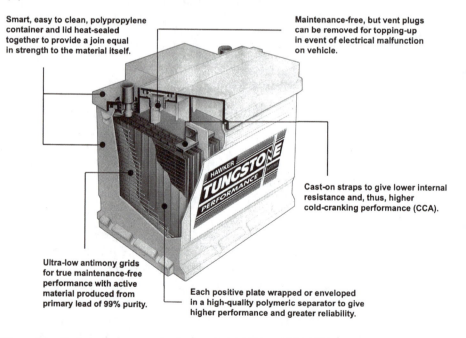

Cast-on straps to give lower internal
resistance and, thus, higher
cold-cranking performance (CCA).

Ultra-low antimony grids
for true maintenance-free
performance with active
material produced from
primary lead of 99% purity.

Each positive plate wrapped or enveloped
in a high-quality polymeric separator to give
higher performance and greater reliability.

Figure 8.2   *Design of an automotive battery in* (a) *1930s and* (b) *1990s*
[By courtesy of (a) Research Studies Press Ltd and (b) Tungstone Batteries
Ltd.]

and tellurium. These agents enhance the stiffness, the tensile strength and the corrosion resistance of the grid alloys. Castability is improved by the addition of 0.1 wt.% Sn which increases the fluidity of lead alloys. Low-antimony alloys, as used for grids in 'low-maintenance' automotive batteries, contain ~ 2 wt.% Sb.

Today, many automotive batteries are of the so-called 'maintenance-free' variety in which the grids are constructed of lead–calcium alloy (up to 0.1 wt.% Ca), often with the addition of tin (0.3 to 0.7 wt.% Sn). On quenching the grids during manufacture, a dispersion of the compound Pb_3Ca forms around the pure-lead grains and results in precipitation hardening. These alloys exhibit a higher hydrogen overpotential than antimonial types so that water loss during charging is largely eliminated, provided that the top-of-charge voltage is correctly controlled. Also, the alloys have improved electrical conductivity, but are substantially weaker than their lead–antimony counterparts. Batteries with lead–calcium grids are normally used in 'float' conditions (*e.g.* automotive or stand-by applications). The batteries are unsuited to regular deep discharge as the cycle-life under such duty is quite limited. Explanations advanced for this are complex. The dominant failure mechanisms are attributed to degradation of the positive plate and involve the build-up and cracking of corrosion layers of substantial electrical resistance on the lead–calcium grid and structural changes in the active material.

8.3.2 Plate Design and Manufacture

Traditionally, grids are cast from molten alloy, either singly or in pairs joined by their lugs, and are quench-cooled. Automated casting machinery is quite sophisticated and much development work has gone into optimizing the casting process, to produce defect-free grids of different alloy compositions. Liquid lead alloy is gravity-fed into the mould where it cools rapidly and then is ejected as a solid grid. With modern casting machinery, the overall process takes only 3 to 4 s, and 15 to 20 pairs of grids are produced each minute. Pairs of positive and negative double-grids, as cast, are shown in Figure 8.3(a).

Over the years, there has been a substantial reduction in the thickness of grids, from more than 2 mm in the 1960s to about 0.8 mm today. Several factors have combined to make this reduction both possible and acceptable, notably, improved casting technology and improved charge-control systems. Overcharge used to cause severe corrosion of the positive grid, which resulted in structural weakness and ultimate failure. Thick grids were therefore used to ensure adequate battery life. Better voltage control restricts the extent of corrosion and, together with the

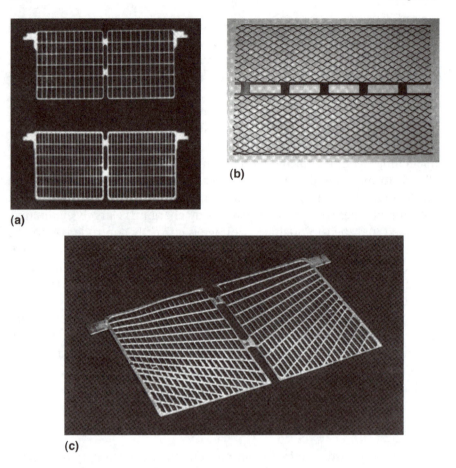

Figure 8.3 *Grid designs: (a) as-cast positive and negative double-grids (note: in the cell these are turned through 90°); (b) expanded metal; (c) positive grid with radial design*
[By courtesy of (a), (c) Research Studies Press Ltd and (b) Batteries International]

improved defect-free castings now available, thinner grids are practicable. Over the same time period, the average life-expectancy of an automotive battery has increased from around two years to four years or more.

There are alternatives to the gravity-casting of grids. A continuous casting process has been developed and employs a rotating, water-cooled drum on which is engraved a continuous, multiple impression of the grid. The liquid alloy is introduced to the drum *via* a 'casting shoe' located directly over the surface of the drum. Another procedure is to start with a continuous reel of alloy foil which is then slit and expanded to form a diamond mesh structure which is suitable for pasting [Figure 8.3(b)].

Expanded-metal technology is particularly suitable for the lead–calcium alloys used in maintenance-free automotive batteries.

Much science has gone into determining the best composition and properties of lead alloys, and a comparable amount of technology into the design and manufacture of grids. As regards design, there is considerable scope for optimizing the configuration. The current-collecting lug is inevitably located on the top edge of the plate, usually at one corner. As current is collected from the electrochemical reaction in the paste spread uniformly over the entire area of the electrode, it follows that the current density increases the nearer the approach to the lug. Therefore, to minimize the internal resistance of the cell, the grid conductance should increase towards the lug. This may be achieved by increasing the cross-section of the ribs of the grid in this area, or by adding ribs angled down from the vicinity of the lug, the so-called 'radial design' of grid [Figure 8.3(c)]. Computer optimization of grid design is valuable, although ultimately the design has to be capable of being manufactured economically and has to withstand the mechanical stresses and vibrations to which it is subjected in the battery.

8.3.3   Electrode Composition and Pasting

The active material is made by first reacting lead ingots with air in a ball mill, or molten lead with air in a furnace (the 'Barton-pot' method). The resulting powder – known as 'leady oxide' – is composed of lead monoxide (PbO) and unreacted lead particles ('free-lead'). In the next stage of plate manufacture, the leady oxide is made into a paste with sulfuric acid solution. During paste-making, a significant proportion of the leady oxide converts into various basic lead sulfates. These compounds serve to consolidate and strengthen the paste, a process which may be likened to the setting of cement. Certain minor additives are included in the negative-plate mix, *i.e.* barium sulfate, lignosulfonates, and carbon black. These additives, collectively known as 'expanders', serve to improve the low-temperature performance and extend the cycle-life of the battery. The expanders prevent individual crystals of lead from growing and combining into a dense structure with low surface-area and, therefore, low electrical capacity. Sometimes fibres and/or binders are added to the paste to give it strength. The pasting operation consists of feeding paste from an overhead hopper on to cast grids, or on to continuous expanded-metal grids, that pass horizontally underneath.

After pasting, the plates are flash-dried and then 'cured'. Curing consists of putting the stacked plates in an enclosure of controlled temperature and humidity for up to 72 h. This treatment allows the development of basic lead sulfates and the oxidation of free lead in the active mass to proceed to completion. These processes serve to harden and strengthen the active mass. Finally, the plates are mounted in the battery, welded into groups of like polarity, and 'formed' by charging electrochemically to produce lead dioxide at the positives and spongy lead metal at the negatives.

8.3.4   Separators

Porous separators, which are placed between the positive and negative plates, allow current to pass *via* ionic conduction through the pores. Although a vital component of the battery, they do contribute significantly to its internal resistance. Originally, separators were made of acid-resistant wood (fabricated as thin sheets), cellulose (in paper form), hard rubber, or resin-impregnated glass-fibre mat. Modern separators usually consist of a microporous plastic material [polyethylene or poly(vinyl chloride)] with an inorganic filler (silica). The separator may contain three times as much silica, by weight, as polymer. The pore volume and pore-size distribution are tightly controlled in the manufacturing process. Ideally, the separator should have $\sim 60\%$ porosity with a uniform pore size of $< 1~\mu$m. The high porosity is needed to minimize the resistance, and the small pore size to limit dendrite penetration. The separator has longitudinal ribs to hold it away from the positive electrode where water is formed on discharge. This ribbed design facilitates diffusional mixing of the electrolyte solution. Much development work has gone into the design of modern separators and their manufacture is a sophisticated industry. A typical plant produces tens of millions of square metres of separators per annum. Sometimes the separator is in the form of an envelope, sealed on three sides, which encloses either the negative or positive plate; this eliminates dendritic growth which tends to cause short-circuits with leaf separators.

8.3.5   Automotive Battery Performance

The specifications for an automotive battery are quite straightforward. First, the capacity (in Ah) should be appropriate to the size of engine to be started and whether it operates on petrol or diesel. After that, the owner is primarily interested in 'fitness for purpose' (*i.e.* reliability and life) and in cost. 'Fitness for purpose' signifies the ability to supply the high current needed for engine starting under all conditions likely to be encountered by the vehicle. This is essentially a matter of ambient temperature in the engine bay: starting will be difficult below $-20\,^\circ$C and ideally lead–acid batteries should not be heated above $+50°C$, although the temperature under the hood of a vehicle may sometimes exceed this upper limit in summer.

'Battery reliability' implies freedom from sudden and unexpected failure without warning. Occasionally this happens, either from an internal short-circuit, or from the mechanical failure of an internal component caused, for example, by brittle fracture or corrosion. Although such faults are likely to be restricted to one cell, the loss of 2 V would make the battery virtually useless.

'Battery life' is defined as being at an end when the power output is inadequate to start the engine or when it cannot be sustained for a sufficient time. Throughout the life of the battery, progressive degradation processes are taking place, particularly at the positive plate. These include: shedding of active material; corrosion of the grids, busbars and terminals (the so-called 'top-lead'); irreversible formation and accumulation of inactive lead sulfate. Cumulatively, such effects result finally in a need to change the battery. Just when this becomes necessary is a subjective judgment, dependent upon the driver and the driving conditions. Often, it is the first cold morning of winter which demonstrates a weak battery. Battery life is shortened considerably if the duty involves regular deep discharge, or if the charge regulator is set at too high a voltage. For a 12 V battery, the charge voltage should not exceed 14.4 to 14.8 V; the limit is determined by the choice of positive grid alloy, *i.e.* lead–antimony or lead–calcium–tin. Remarkably, on fitting a replacement battery, it is often not appreciated that the voltage regulator may have to be adjusted accordingly.

Despite all of the above problems, the modern automotive battery is highly reliable and is generally long-lived if treated correctly and not abused. Lives of three to four years are quite normal when in daily use, and much longer lives have been reported. Moreover, battery cost, in real terms, has declined markedly over the years as performance has been improved, the mass reduced, and the need for maintenance almost eliminated. In truth, the automotive battery has been a great success story, vital to the performance of the modern motor vehicle.

Marked changes in the design of automotive batteries are imminent. Increased power demands on automotive electrical systems to enhance vehicle performance, convenience and safety – in addition to the industry's need to improve fuel economy and reduce emissions – are making present 12/14 V systems inadequate. Accordingly, car manufacturers are planning to elevate the system voltage to the maximum safe level of 42 V and the nominal battery voltage to 36 V. Apart from increasing electrical power and efficiency, the higher voltage will accommodate additional loads and permit the implementation of new technologies. Features to be incorporated in future vehicles include: electric air-conditioning; high-power water pumps; integrated starter/generators; power steering; steer-by-wire; brake-by-wire; ride control systems; mobile office, navigation and collision-avoidance devices; automatic engine stop at idle. Higher voltage (and, therefore, lower current) will also trim weight from vehicles by allowing the use of thinner and lighter wiring and cables, as well as smaller connectors. This is also expected to bring cost savings.

While the efficiency of most electrical appliances on board vehicles is enhanced by elevated voltage, some devices such as incandescent lamps and computers prefer low voltage. As a result, two battery options are presently being evaluated: (i) a single dual-output battery, *i.e.* a universal 36 V unit with a d.c.–d.c. conversion capability for specified 12 V loads; (ii) a dual-battery, *i.e.* a 12 V, high-capacity, cycling unit for auxiliaries and a 36 V, high-power, unit for engine starting. Option (i) appears to be less desirable as preliminary designs based on conventional lead–acid technology indicate a battery which is heavy, bulky and expensive, with less than optimum performance characteristics. By contrast, option (ii) allows greater flexibility in battery design in that each unit can be optimized for its specific assignment. A dual-battery also provides for assured starts, since the 36 V unit would not be deeply discharged. The 12 V unit is likely to be an extension of the present lead–acid automotive battery – quite possibly a valve-regulated design (see Section 8.7) – but its 36 V partner must embody significant technological innovation to meet stringent size, weight, performance and cost targets. In the long-term, it is foreseen that a battery technology other than lead–acid, *e.g.* lithium-ion or lithium–polymer (see Sections 10.4 and 10.5, Chapter 10), will be used for the high-voltage power source, especially when vehicles complete the transition to a single 42 V system.

8.4   LEISURE BATTERIES

There are applications in the leisure market where deep cycling is required and for which the automotive battery is unsuitable on account of its short life under such conditions. Examples are traction batteries for golf carts, and batteries for 'domestic duties' in caravans and boats. The latter duties include internal lighting and the operation of radio and television sets, refrigerators, water pumps, shower exhaust pumps, electric lavatories, *etc.* In the case of boats, power is required also for bilge pumps, navigation lights, and instrument panels. The battery is recharged from the engine alternator during the day and may experience quite a deep discharge overnight; the refrigerator alone may take 6 to 10 A (*i.e.* 70

to 120 W). The ideal battery for these applications is the tubular traction battery (see Section 8.6), but it is rarely used because it is expensive and heavy. Instead, the manufacturers have developed 'leisure batteries' which are a special form of pasted-plate battery.

Generally, leisure batteries employ grids made from lead–2 wt.% antimony alloy and so require periodic make-up of water. Often, the grids are thicker than those used in automotive batteries. This is to allow for increased corrosion and to provide a robust plate which can withstand the mechanical stresses induced by deep discharge (as noted in Chapter 4, stresses arise because the volume of the discharge product, lead sulfate, is greater than that of the precursor lead dioxide). The employment of thicker plates is possible because the high currents associated with engine starting are not required. Shedding of positive active-material is moderated by wrapping the plates in a glass mat and packing the cell assembly tightly into the container to enhance the 'plate-group pressure'. Leisure batteries also have an enhanced space under the plates to accommodate greater amounts of shed material without promoting short-circuits. The details of cell design vary according to the manufacturer. Many leisure batteries have integral handles for carrying. To specify the size of leisure battery required for a particular application, it is necessary to total up the number of ampere-hours likely to be consumed daily by each of the appliances, and then add a margin for prudence. Typically, a residential diesel-engined boat will have an automotive battery for engine starting and two or three batteries (80 to 110 Ah each) connected in parallel for domestic duties.

So-called 'semi-traction' batteries are similar to leisure batteries and are used to power small electric vehicles (floor sweepers, fork-lift trucks, golf carts, pallet trucks, wheelchairs, *etc.*). They are available in a wide range of capacities (30 to 180 Ah, $C_5/5$ rate) and, to a large extent, are interchangeable with tubular traction batteries. Essentially, semi-traction batteries are cheaper than tubular equivalents, but not so durable. The batteries are therefore the preferred choice for lighter duties.

8.5 STATIONARY BATTERIES / UNINTERRUPTIBLE POWER SUPPLIES

There are many applications for stationary (stand-by) batteries, as outlined in Section 1.2, Chapter 1. Historically, the requirement for stand-by power arose first in telephone exchanges and hospital operating theatres, but has now expanded into other areas of modern technology where uninterruptible power supplies (UPSs) are essential. These areas include: central computing facilities; power generation and transmission (including such critical situations as nuclear power control rooms); switch tripping; emergency lighting; gas-turbine or diesel-engine starting. On the railways, stand-by batteries are used to back-up track signalling and carriage lighting systems. Wherever the loss of mains power would be critical, it is necessary to install an emergency stand-by battery which would take over seamlessly. Often, there is also a stationary diesel generator installed, in which case the battery has only to provide power for as long as it takes to start the generator. This period is likely to be less than 15 min and, therefore, a battery which is capable of providing a high-rate discharge is required.

The traditional stand-by battery is based on the so-called 'Planté design'. The fundamental difference between Planté cells and all other lead–acid cells lies in the use of a single solid casting of pure lead for the positive plate. The active material is formed electrochemically on the positive plate, so no mechanical bonding of separately prepared paste is required. It is this which makes Planté cells the most reliable of all lead–acid cells. Early Planté cells had a thick solid plate of lead for the positive electrode. Over the years, the design has moved towards the use of a thinner plate which has regular vertical fins perpendicular to its plane (Figure 8.4). The effect of these fins is to increase substantially the effective surface area of the plate and thus improve the stored capacity per unit mass and unit volume. Even so, the capacity of the Planté cell is poor compared with other types of lead–acid battery. The virtues of the cell are its reliability and long life on float duties.

The negative plates of Planté cells are conventional pasted types which use low-antimony lead alloy. These plates offer a service-life similar to that of the positives. During service, active material is gradually lost from the positive plate and falls to the bottom of the cell as a sediment. Active material is continuously reformed at the surface of the plate so that the cell retains a constant performance. The space beneath the plates is

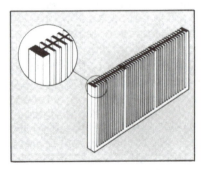

Figure 8.4   *Positive-plate construction in a Planté cell*
            (By courtesy of Tungstone Batteries Ltd)

adequate to accommodate the sediment expected over 25 years of service. The cell is housed in a transparent plastic container so that it is possible to observe the accumulation of sediment directly and estimate whether end-of-life is imminent. A selection of small Planté cells manufactured by Tungstone Batteries Ltd. is presented in Figure 8.5. A very large UPS installation of the type used in a power station or other major facility is shown in Figure 8.6.

When Planté cells are held on float charge at 2.25 V per cell, and at a temperature of 20 to 25 °C, they should have a life in excess of 20 years. Maintenance (water top-up) is conducted annually. Because the cells are vented, released gas bubbles do tend to entrap a fine sulfuric acid spray which can cause corrosion of external equipment when the cells are housed in an enclosed room.

Some manufacturers market a version of the tubular battery (Section 8.6) for stand-by power applications. The design is available as single cells of varying size (200 to 3400 Ah), or as 6 V or 12 V monoblocs (Figure 8.7). The cells are fitted with flame-arresting vents. Other manufacturers market a version of the flat, pasted-plate cell with low-antimony (2 wt.% Sb) positive grids for stand-by power applications. These cells are cheaper than their Planté and tubular counterparts, but are likely to have a shorter life.

Around 10 years ago, work was undertaken in France to develop tubular batteries with lead–calcium alloy spines for UPS applications in

Figure 8.5 *Small Planté cells*
(By courtesy of Tungstone Batteries Ltd)

Figure 8.6   *Large UPS battery room*
(By courtesy of Tungstone Batteries Ltd)

Figure 8.7   *Vented tubular monoblocs (6 V and 12 V) manufactured by Varta Batterie AG*
for stand-by applications
(By courtesy of Varta Batterie AG)

nuclear power stations. It was claimed that the float current was low and remained low for much longer than with the Planté cells. Accordingly, the new batteries required less maintenance. Recently, VRLA batteries (Section 8.7) have been adopted for some UPS applications. These have the advantages of no water maintenance and no acid spray. Moreover, VRLA batteries can be operated horizontally, which allows multi-stacking to provide a smaller footprint and easier access for voltage measurements. One of the largest VRLA batteries for UPS duties is the unit installed by GNB Technologies at its lead-recycling plant in Vernon,

California. When utility power is interrupted or lost, the battery system can provide up to 5 MW in support of all the plant loads.

In summary, a wide variety of lead–acid designs (Planté, tubular, flat-plate, valve-regulated) are marketed for UPS applications and these are available as either cells or monoblocs in a wide range of capacities. In making a choice for a particular application, it is necessary to consider all the relevant factors, *e.g.* duty cycle, desired life, frequency of maintenance, float current, release of gases and acid mist, cost.

8.6 TUBULAR TRACTION BATTERIES

In a tubular cell, the positive plate is constructed from a series of vertical lead alloy spines or 'fingers' that essentially resemble a comb. These act as current-collectors and are inserted into tubes made from woven, braided or felted fibres of glass or polyester. The tubes may be either mounted individually or joined together in a row (the 'gauntlet' design) with spacing equal to that between the spines. The tubes are sealed at the base with plastic caps which are mounted on a common bar. The active material is packed into the annulus between the spine and the tube wall. This is shown schematically in Figure 8.8 where a comparison is made between the structure of flat pasted plates and tubular plates. With the tubular design, it is not possible to shed active material, except in cases of severe battery misuse where splitting of the tubes may occur.

Tubular traction batteries have conventional pasted plates as negative electrodes and conventional separators. A cut-away of a 2 V cell is shown

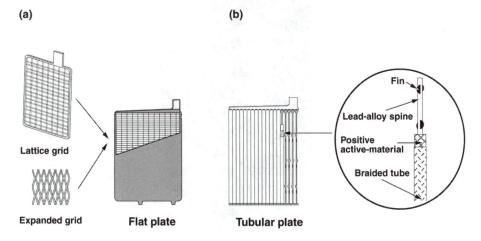

(a) **(b)**

Lattice grid

Expanded grid   **Flat plate**   **Tubular plate**

Fin

Lead-alloy spine

Positive active-material

Braided tube

Figure 8.8 *Lead–acid batteries with* (a) *flat plates and* (b) *tubular plates*
(By courtesy of Research Studies Press Ltd)

in Figure 8.9; the tubular positive plate and the grid of the negative plate are clearly seen.

Lead–antimony alloy is used for the spines of the positive plate. For this reason, the batteries are prone to lose water and require regular servicing. With a large traction battery, this is a time-consuming chore and so several manufacturers have devised an automatic watering system. Basically this consists of a water reservoir under pressure which feeds water through transparent plastic tubing to filler plugs at the top of each cell. As the electrolyte level rises, a float shuts off the water supply to each cell individually. It is claimed that automatic watering not only saves labour costs, but also prolongs battery life and performance by ensuring topping-up to the correct level.

Individual, 2 V, traction cells range in capacity from less than 100 Ah up to 1800 Ah. The smallest weigh ~ 7 kg and the largest ~ 100 kg. As an approximation, the capacity per unit mass is ~ 13 Ah kg^{-1} for the smaller cells and ~ 18 Ah kg^{-1} for the largest sizes. All cells are built to a standard width, with variable length and height. The British Standard width is 159 mm and the German DIN standard is 198 mm. These single

Figure 8.9 *Anatomy of a 2 V tubular traction cell*
(By courtesy of Batteries International)

cells, which are used for larger vehicles, are joined together in series to give the desired voltage. For smaller traction applications, such as wheelchairs or golf carts, 6 V and 12 V monoblocs are available in various sizes up to ~ 150 Ah.

To minimize the effect of acid stratification in the larger cells, some manufacturers have resorted to induced electrolyte circulation. One such method involves bubbling air through the sulfuric acid. A diaphragm pump blows air down the inner of two concentric tubes which reaches to near the bottom of the cell. The air then rises through the outer tube and, thereby, promotes a flow of electrolyte in the cell. Many advantages are claimed for this procedure, namely:

- no electrolyte or temperature stratification;
- uniform charge-acceptance across the area of the plates;
- lower temperature rise during charging;
- reduced charging times – faster charging is possible;
- minimized gassing;
- reduced maintenance and water consumption (up to 75%);
- higher performance and longer battery service-life;
- energy savings of up to 20%.

If these claims are all substantiated it seems surprising that electrolyte circulation is not adopted universally for large traction cells. The answer may lie in the added complexity, increased volume and extra cost of the system.

Some of the largest tubular batteries are those employed in submarines to supply all the power requirements of the boat when submerged. Usually, there are two independent battery banks, one along the starboard side and the other along the port side of the boat, so as to balance the mass. Individual cells are of height 1.0 to 1.4 m and weigh from 300 to 700 kg. The weight depends upon the required capacity, which lies in the range 5000 to 11 000 Ah. Cells of this size require air bubblers to minimize acid stratification. The cells have to be hoisted aboard by crane, lowered through the conning-tower, manoeuvred along the confined inner-deck space of the vessel, and lowered into a special battery housing below the deck. Quite a feat! Because of height limitations, a removable deck is sometimes fitted to permit access to the cells for maintenance.

The operational specification for submarine batteries is quite specific. Gassing has to be reduced to a minimum to avoid build-up of hydrogen in the atmosphere and to restrict the maintenance required. Long life, reliability and safety are paramount, as also is high power output (up to several MW) for high speed when pursuing or escaping from the enemy.

High charge-acceptability is important to reduce the time the vessel must remain exposed on the surface when recharging. In normal operation, the battery is not discharged to more than $\sim 30\%$ DoD, as no submarine commander wishes to run short of power when submerged. The partial-discharge duty also prolongs the service-life of the battery. Interestingly, nuclear submarines have cells of comparable size to those in conventional submarines, but fewer cells in the battery. This is a stand-by battery which is required in an emergency to shut down the reactor and to provide electrical power for other essential equipment thereafter.

8.7   VALVE-REGULATED BATTERIES

Until recent years, all lead–acid batteries were vented and therefore had to be used in the upright position. This was (and still is) a serious limitation for small batteries in portable applications and for larger batteries in, for example, military aircraft where aerobatic manoeuvres would cause acid spillage. Many attempts were made to develop hermetically sealed batteries, but these always foundered on the problem of build-up of an explosive mixture of hydrogen and oxygen on approaching the top-of-charge and during overcharge. Most early attempts to solve this problem were based on the use of platinum metal catalysts to facilitate the hydrogen–oxygen recombination reaction. Generally, the platinum metal was supported on the surface of a porous pellet of finely-divided alumina, silica or carbon which was located at the top of the cell, above the acid. Initially, catalytic recombination works well, but after a while the pores of the pellet become filled with water, which restricts further access of gas. No fully satisfactory catalytic device has been developed for use in lead–acid batteries of the flooded-electrolyte design.

The alternative approach to developing a 'sealed' battery is to recombine the oxygen at the negative electrode *via* the so-called 'internal oxygen cycle', as follows.

At the positive electrode:

$$H_2O \longrightarrow 2H^+ + \tfrac{1}{2}O_2 + 2e^- \tag{8.1}$$

At the negative electrode:

$$Pb + \tfrac{1}{2}O_2 + H_2SO_4 \longrightarrow PbSO_4 + H_2O \tag{8.2}$$

$$PbSO_4 + 2H^+ + 2e^- \longrightarrow Pb + H_2SO_4 \tag{8.3}$$

In this cycle, the oxygen discharges chemically at the negative electrode [equation (8.2)]. This shifts the potential to a more positive value and thus decreases the rate of hydrogen evolution to a low level (as noted earlier, a hydrogen-recombination cycle is kinetically impossible). Since the negative electrode is simultaneously on charge, the discharge product is immediately reduced electrochemically to lead [equation (8.3)] and the chemical balance of the cell is restored.

For the cell design to be practical, it is necessary to provide a path whereby oxygen liberated at the positive electrode can reach the negative electrode. This can be achieved in the so-called 'valve-regulated lead–acid' (VRLA) cell. The cell has an 'electrolyte-starved' design in which the sulfuric acid is immobilized in the separator and the active materials, and sufficient empty porosity is left for oxygen to diffuse through the separator to the negative plate. Electrolyte-starved cells stand in marked contrast to the conventional, flooded-electrolyte cells in which the free acid covers all the plates and the busbars.

There are two methods of immobilizing the electrolyte between the plates in VRLA cells:

- *The gel battery*. Sulfuric acid is mixed with very finely-divided, high surface-area, silica ('fumed silica'). This forms a viscous solution which develops into a gel on standing. Oxygen transfer occurs through fissures in the gel that arise during the early stages of battery life as a result of partial drying out and shrinkage. This process was introduced in the mid-1960s in Germany by Sonnenschein GmbH (now part of the Exide Group), and has been used to produce gel cells which are marketed under the Dryfit® brandname.
- *The AGM battery*. AGM stands for Absorptive Glass Micro-fibre, from which the separator in this battery is made. The AGM is usually saturated with acid to between 90 and 95% of its absorptive capacity. The remaining pore volume is available for the passage of oxygen to the negative plate. The AGM separators must be in close contact with the plates to minimize internal resistance by allowing fast transfer of ions during both charge and discharge. To achieve sufficient contact, the plate group is compressed *via* a tight fit with the container. The first commercially successful cell of this type was the Cyclon™ cell. This was developed in the USA in 1971 by the Gates Corporation and is now manufactured by Hawker Energy Products, Inc.

The relative advantages and disadvantages of gel and AGM types of VRLA battery with respect to flooded-electrolyte counterparts are listed in Table 8.2.

Table 8.2 *Advantages and disadvantages of VRLA batteries*

Advantages	Disadvantages
• No water addition	• Careful charging required
• No acid spillage	• Thermal management is more critical
• Negligible acid fumes	(especially for AGM type)
• Easy transportation	• Significant variation in top-of-charge
• No special ventilation requirements	voltages
• Operable in any orientation	• Increase in overcharge required at
• Smaller footprint (stacked batteries)	elevated temperatures
• Negligible acid stratification (gel type)	• Deep cycle-life often inferior
• Less overcharge required at room	• Cannot measure relative density
temperature	• Not available in dry-charged state
• Good high-rate, discharge capacity	• Shelf-life of 2 years maximum
(AGM type only)	

The Cyclon™ cell (Figure 8.10) is a cylindrical, AGM cell which is manufactured in D-, X- and BC-sizes with capacities ($C_{10}/10$ rate) of 2.5, 5.0 and 25 Ah, respectively. As these are sealed cells, the smallest of the three may be used as a rechargeable replacement for a D-size primary cell, albeit at 2 V rather than at 1.5 V. Six-volt monoblocs are also available. Both the positive and the negative plate of the Cyclon™ cell are made of pure lead–tin alloy and are quite thin. (The role of tin is to enhance the mechanical properties of the alloy, to increase the corrosion resistance, and to improve the electrical conductivity of the plate | active-material interface.) Other manufacturers of VRLA cells prefer to use lead–calcium–tin alloy, although these cells are more suited to float duties than to cycling. The grids are pasted with lead oxides, separated by a layer of AGM, and then spirally wound. Lead busbars are welded to the exposed plate lugs before the assembly is inserted in the container. Sulfuric acid is added through the vent hole which is then sealed with the pressure-release valve. As these cells are deemed to be totally safe, they are exempt from most shipping regulations, unlike vented cells.

By virtue of their very low internal resistance, Cyclon™ cells have very high pulse-power capabilities. The maximum current from 2.5 and 5.0 Ah cells is 65 A, and that from the 25 Ah cell is 250 A. The cells also have very good low-temperature performance down to at least $-20\,°C$, but do not retain their capacity well at elevated temperature. For instance, on standing, 20% loss of capacity can occur in a year at 20 °C, but in only six weeks at 40 °C. Nevertheless, the cells may be charged or held on float at temperatures up to 65 °C. Provided careful charge control is exercised, Cyclon™ cells have a cycle-life of ~ 300 cycles for deep discharge and up to 2000 cycles for shallow discharge. They are also very suitable for float

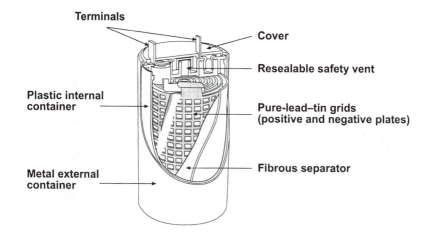

Figure 8.10 *Cyclon*^{TM} *cell*
(By courtesy of Hawker Energy Products Inc.)

duties (*e.g.* memory support) in mains-operated electronic equipment.

The Bolder cell is a novel type of spirally-wound VRLA (AGM) design. The cell uses a pair of very thin (0.21 mm) plates made from ultra-thin (0.05 mm) lead–tin alloy foil. Each plate is coated on both sides with a layer (0.08 mm) of active material, so as to leave a narrow strip of uncoated foil along one edge. The positive and negative plates are offset from each other and then wound together with a microglass separator between them. The result is a spiral assembly with the uncoated strip of positive-plate foil at one end of the cell and the uncoated strip of negative-plate foil at the other end. Lead end-caps are then cast on to each end of the spiral to make electrical contact with the entire length of the positive and negative electrodes, respectively (Figure 8.11). The casting serves as a seal, a current-collector, a vent, and a manifold for electrolyte distribution. The use of opposite-end current collection, with continuous contact between the plugs and the foils, results in uniform current distribution over the electrode surfaces. This, together with the high surface area of the plates, gives a cell of exceptionally low internal resistance.

It is well known that spirally-wound cells are of higher power output and lower capacity than bobbin or plate cells. The Bolder concept carries this to its logical conclusion by using very thin plates and configuring them into a cell of very low resistance. The result is that a small 1.2 Ah cell, weighing only 77 g, may be fully discharged at the 30 A (30C) rate. This corresponds to a specific power of almost 800 W kg^{-1}. In fact, these cells will deliver more than 800 W kg^{-1} to 70% DoD, and will even give 400 W kg^{-1} at 90% DoD. Although the capacity of Bolder cells is

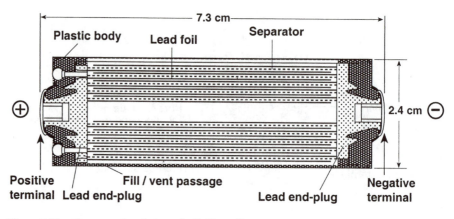

Figure 8.11   *Cross-sectional view of a Bolder cell*
                  (By courtesy of Elsevier Science)

necessarily limited, in portable tool applications they are said to perform better than nickel–cadmium cells of equivalent volume. Bolder technology is so powerful that a pack of six small cells, weighing only 0.5 kg, is able to start a car engine numerous times before having to be recharged. Moreover, the subsequent recharge can be accomplished in only ten minutes.

Larger, flat-plate VRLA batteries are manufactured as single cells of capacity up to 3000 Ah, or as 6 V or 12 V monoblocs in sizes up to about 500 Ah. As they are sealed and require no maintenance, they may be rack-mounted in layers, one above the other, to save space. For float applications at 20 °C it is recommended that the voltage applied to the terminals shall be in the range 2.27 to 2.29 V per cell. As this value varies sharply with temperature, charge control is critical. These batteries are subject to the same advantages and disadvantages as spirally-wound cells, and are marketed as being suitable for a wide variety of duties which range from stand-by power for telecommunications systems to motive power for electric vehicles. In general, they are not consumer batteries and the professional user should make a careful technical evaluation of their attributes and drawbacks for the intended application in comparison with other types of battery available. In the technical literature, there has been much debate over the relative merits of gel and AGM types of VRLA battery in terms of electrical performance and service-life in different applications, and the issue is by no means yet resolved.

8.8 MARKETS

In 1999, the world market for automotive and leisure batteries was estimated as 310 million monoblocs per annum. This market is growing as the world fleet of motor vehicles expands, and as more and more motorized equipment comes into use in industry, agriculture, and the home. Examples are bulldozers, farm machinery, tractors, motor boats, large d.c. generators, and grass-cutting machines – all of which require a starter battery.

Industrial batteries are larger and heavier than automotive batteries and this sector encompasses both stationary and traction batteries. A 12 V unit may weigh anything from 35 kg up to several tons. The world market for industrial batteries in 1999 was estimated at 48 million units per annum.

Finally, we have small sealed lead–acid batteries for use by professional electronics engineers and by consumers. In 1999, this world market was 60 million units per annum.

In terms of value, the present world market for lead–acid batteries is ~ US$10 billion per annum. By comparison, the US market in 1997 for all types of rechargeable battery was US$5.6 billion, with growth projected to reach US$7.9 billion by 2003.

Chapter 9

Alkaline Batteries

Because many common metals are not stable in acid solution, there is more scope for developing aqueous batteries with neutral or alkaline electrolytes. The choice of potassium hydroxide (KOH), rather than the cheaper sodium hydroxide (NaOH), for use in alkaline batteries is determined by its higher electrical conductance. The comparative conductance data at 25 °C are shown in Figure 9.1. The conductance of 6 to 8 M KOH solution is comparable with that of battery sulfuric acid (see Section 8.2, Chapter 8), whereas that of 6 to 8 M NaOH solution is only about two-thirds as high. The use of sodium hydroxide would therefore result in a significant increase in the internal resistance of the battery.

The disadvantage of using a concentrated solution of potassium hydroxide is that it shows a marked tendency to creep and to seep from seals, whereas sodium hydroxide is less difficult to contain. Nevertheless, the conductance advantage outweighs this problem and potassium hy-

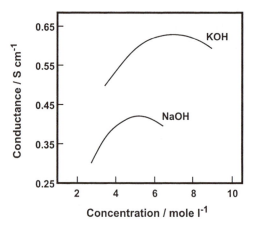

Figure 9.1 *Conductance of potassium and sodium hydroxide solutions at 25 °C.*

droxide is the usual electrolyte. On some occasions, sodium hydroxide is employed in small cells where operation at low temperatures and/or high current drains is not required, and where long service-life without leakage is essential. Such conditions apply to cells soldered on printed-circuit boards and to button cells used for memory back-up.

Rechargeable batteries with alkaline electrolytes employ one of three metals (cadmium, iron, or zinc) as the negative electrode, and an oxide of a metal in a higher valence state as the positive. The latter electrode is either manganese(IV) oxide, nickel(III) oxide, or silver(II) oxide. Cells with electrodes based on manganese oxide or silver oxide are usually assembled in the charged state using, respectively, MnO_2 or AgO. By contrast, cells with nickel oxide positives are assembled in the discharged state and use nickel hydroxide, $Ni(OH)_2$, as the active material which is then converted by electrical charging into a compound of nickel which is in a higher valence state. The charged compound is generally taken to be trivalent nickel oxyhydroxide, NiOOH. This is an imprecise description, however, as the fully-charged electrode also contains some quadrivalent nickel. Following convention, it is convenient to refer to the positive electrode simply as the 'nickel oxide electrode'.

There are also rechargeable alkaline batteries in which the metal of the negative electrode is replaced by hydrogen (*e.g.* the hydrogen–nickel-oxide battery, usually termed the 'nickel–hydrogen' battery) or the metal oxide of the positive electrode is replaced by oxygen (*e.g.* the zinc–air battery). In effect, such systems are battery–fuel-cell hybrids in that they can be recharged (a battery characteristic) and use a reactant, *viz.* hydrogen or air, which is stored externally to the electrodes (a fuel-cell characteristic). The nickel–metal-hydride battery, which is a relative newcomer, should be classed as a battery because the hydrogen is immobilized in the negative electrode. All types of rechargeable alkaline battery will be described in this chapter.

A word on nomenclature is perhaps timely. Strictly speaking, the convention is to describe a battery by stating the negative electrode first, followed by the positive. Thus, we have 'zinc–manganese-dioxide', 'sodium–sulfur', 'lithium–iron-sulfide', *etc.* An obvious exception is 'lead–lead-dioxide', which is invariably known as 'lead–acid'. In the case of the alkaline batteries, these should correctly be designated 'cadmium–nickel-oxide', 'iron–nickel-oxide', 'zinc–silver-oxide', *etc.*, but they are usually referred to as nickel–cadmium, nickel–iron, silver–zinc, respectively. This is just one more idiosyncrasy of battery lore that the reader new to the science has to accept.

9.1   RECHARGEABLE ALKALINE-MANGANESE CELLS

Alkaline-manganese cells (see Section 5.2, Chapter 5) are normally regarded as primary cells. Indeed, users are warned against attempting to
recharge them as this can lead to the build-up of internal gas pressure and
the possibility of violent rupture of the cell container. Recent research has
shown, however, that it is indeed possible to recharge this class of cell
safely, provided the correct design is used and the charge–discharge
conditions are tightly controlled. Such demonstration has led to a new
type of consumer alkaline cell, the RAM cell (Rechargeable Alkaline-
Manganese cell), which combines the performance of alkaline-manganese
technology with the benefit of reuseability for a limited number of cycles.
Such a system might be termed a 'quasi-secondary battery'.

What have been the advances which have made possible the recharging
of alkaline-manganese cells? First, new separators were developed. These
are highly stable in the strongly alkaline environment and also prevent
the formation of zinc dendrites that otherwise would cause internal
short-circuiting. Measures were also taken to prevent swelling of the
positive electrode. In particular, the manganese oxide electrode was made
rechargeable by limiting the capacity of the zinc electrode so that discharge does not take place beyond transfer of the first electron (*i.e.*
$Mn^{4+} + e^- \rightarrow Mn^{3+}$); this corresponds to a cut-off cell voltage of 0.9 V.
To allow for hydrogen formed by corrosion, catalysts are added to the
positive-electrode mix to facilitate the recombination of hydrogen gas,
and also to provide overcharge protection *via* the oxygen-recombination
cycle. Special equipment has been designed to taper-charge RAM cells to
a maximum of 1.7 V, so as to prevent gassing.

Alkaline-manganese cells have a considerably higher capacity than
conventional rechargeable cells of similar size (Table 9.1). Although not
all of this capacity is available on recharge, the incentive to recharge
alkaline-manganese cells is still substantial, particularly when the higher
voltage and the lower cost are taken into account.

When discharged to the one-electron step, RAM cells are capable of
providing at least ten times the service hours of single-use alkaline cells,

Table 9.1 *Capacity and stored energy of AA-size consumer cells*

	Capacity (Ah)	*Voltage* (V)	*Stored energy* (Wh)
Alkaline-manganese primary cell	2.0–3.0	1.5	3.0–4.5
Nickel–cadmium	0.5–1.0	1.2	0.6–1.2
Nickel–metal-hydride	1.0–1.5	1.2	1.2–1.8

while if discharge is restricted to about 20% DoD many more hours of service may be obtained. On cycling with a shallow discharge, a cumulative charge of up to 200 Ah is possible from a 2.5 Ah cell. Bearing in mind that these cells are cheaper than nickel–cadmium, and do not suffer from the memory effect (see next section), or from disposal problems with toxic cadmium, the attraction of RAM cells is obvious, even though their cycle-life is less than nickel–cadmium. The cells are also more suitable for high-temperature applications, where nickel–cadmium does not excel, and they have very low self-discharge rates. Typically, they may be stored for up to five years before use, whereas nickel–cadmium cells lose their charge in a matter of months on standing. There are two other attractive features of RAM cells, namely, they have an output of 1.5 V which allows immediate interchange with primary cells fitted to existing circuitry, and they come ready-charged straight from the packet.

The RAM technology has now been commercialized and cells (bobbin design) are available up to the D-size, along with custom-designed chargers. A well-known manufacturer is Rayovac, Inc. Many of the cells are produced in China where a huge manufacturing capability and a strong market have developed. It seems likely that the world-wide market for RAM cells will continue to grow for those who are prepared to take the trouble of recharging their batteries. RAM cells are ideal for discharge at the 2- to 4-h rate and are therefore suitable for applications such as portable compact disc and tape players, laptop computers, hand-held video games, and toys. The maximum power output is much lower than that of either nickel–cadmium or lead–acid Cyclon™ cells, and thus RAM cells are less suitable for high-rate applications.

An interesting demonstration was mounted in Austria, in May 1990, when RAM cells were still in the developmental stage. A pack of 400 AA-size cells were wired in a 20×20 array to give a 30 V battery with a specific energy of more than 60 Wh kg^{-1}. This was used to power an electric bicycle in a race across a 2400 m mountain pass in the Austrian Alps. The bicycle won the race in conditions of freezing rain and thunder storms. The battery provided 600 W continuously (1 A per cell) on the uphill stretches and received regenerative recharging on the downward slopes. Regeneration involves the use of the motor as a generator during hill descent, and this converts the vehicle's potential energy into electrical energy which recharges the battery. Motive power may not be an economic use for RAM cells, but it certainly demonstrated their capability in a dramatic fashion.

9.2 NICKEL OXIDE POSITIVE ELECTRODES

By far the majority of rechargeable alkaline batteries are based upon nickel oxyhydroxide (NiOOH) positive electrodes which, as mentioned earlier, are more conveniently termed 'nickel oxide electrodes'. There are four basic methods for preparing such electrodes, as follows.

Pocket-plate electrodes. This is the traditional method. The positive active-material is contained in flat, rectangular channels (or 'pockets') made from perforated, nickel-plated steel strip (Figure 9.2). The channels are then interlinked horizontally to form a plate. The active material consists of a mixture of nickel hydroxide, $Ni(OH)_2$, and either nickel flake or graphite. The latter component is added to provide a matrix of sufficient electrical conductivity. An unwanted side-effect of graphite addition is that it may oxidize to carbon dioxide during overcharge, especially at high temperatures. This gas then reacts with the electrolyte to form potassium carbonate which increases the internal resistance of the cell. The problem does not arise with sintered electrodes which contain no graphite. Often, a small amount (~ 5 wt.%) of cobalt hydroxide is added to the positive active-material and serves to improve both the capacity and the cycle-life of the cell. In effect, cobalt increases the electrical conductivity of the active material, reduces the electrode swelling, and raises the overpotential at which oxygen is liberated during charging. The disadvantage of adding cobalt is purely economic – it is far more expensive than nickel.

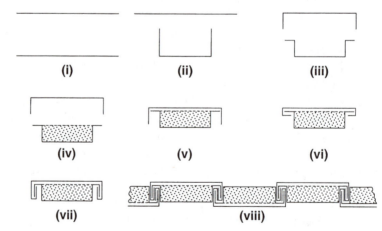

Figure 9.2 *Construction of pocket-plate, nickel oxide electrodes: (i)–(iii) formation of channels in nickel-plated perforated steel strip; (iv)–(vi) filling and crimping to form a long continuous pocket; (vii), (viii) interfacing of filled strips and compression to form final plate*
(By courtesy of Research Studies Press Ltd)

Plastic-bonded electrodes. The active material is mixed with a polymer binder and simply bonded, by rolling or pressing, to the electrode substrate. This electrode has the lowest cost and therefore finds widespread application in small rechargeable cells for consumer products.

Sintered-plate electrodes. The origin of sintered nickel electrodes goes back to World War II when cells based on such electrodes were manufactured in Germany for use in aircraft and rockets. There have been marked improvements since that time. A porous nickel plaque is prepared by lightly compacting nickel powder on to a perforated nickel screen or gauze and then sintering in hydrogen at high temperature. The plaque, which should have $\sim 80\%$ porosity, is charged with active material by chemical or electrochemical impregnation from a bath of nickel and cobalt nitrates. Sintered-plate electrodes have a lower internal resistance than the above types and are capable of delivering higher specific energy and higher specific power. The electrodes are more expensive, however, and their use is confined to batteries of premium grade.

Fibre electrodes. These electrodes have been developed over the past 20 to 30 years and are composed of compressed nickel fibres, or nickel felts or foams. The electrodes are filled with the active material either by applying a paste or slurry, or by chemical or electrochemical impregnation *via* the procedure described above for sintered-plate electrodes.

9.3 NICKEL–IRON BATTERIES

The nickel–iron battery was commercialized during the early years of the 20th century following research and development work by Thomas Edison in the USA (see Section 1.1, Chapter 1). The system comprises nickel oxyhydroxide as the positive electrode, and metallic iron powder as the negative. The electrolyte is concentrated potassium hydroxide which contains added lithium hydroxide to exert a stabilizing effect on the capacity of the positive electrode during charge–discharge cycling. Specifically, the additive minimizes coagulation of the active material. The overall cell reaction is represented by:

$$\text{Fe} + 2\text{NiOOH} + 2\text{H}_2\text{O} \underset{\text{Charge}}{\overset{\text{Discharge}}{\rightleftharpoons}} \text{Fe(OH)}_2 + 2\text{Ni(OH)}_2 \qquad (9.1)$$

$$V^\circ = +1.37\,\text{V}$$

The practical discharge voltage is 1.25 V. Complete discharge of an iron electrode takes place in two steps, first to Fe(OH)_2 and then to Fe(OH)_3. In practice, the second of these steps is not allowed to occur by designing the cells to be 'positive-limited', *i.e.* there is an excess of iron with respect to nickel oxyhydroxide.

The positive and negative electrodes of nickel–iron batteries are of the pocket-plate variety. The pockets are first loaded with nickel hydroxide powder (positive) and iron hydroxide, $Fe(OH)_2$, powder (negative). The respective active materials are then formed electrochemically by charging. Sometimes, copper powder is mixed with the negative material to increase its electrical conductivity. Nickel–iron batteries found early industrial use as traction batteries in fork-lift trucks, mine and railway locomotives, and lanterns. The attraction of nickel–iron batteries is that they have 1.5 to 2 times the specific energy of lead–acid batteries and are particularly good at high discharge rates (Table 9.2). The battery is also noted for its ruggedness and long cycle-life at deep discharge (*e.g.* 2000 cycles at 80% DoD). Compared with lead–acid, however, nickel–iron has several disadvantages, namely:

- inferior performance at low temperatures;
- comparatively high corrosion and self-discharge rates;
- poor overall energy efficiency because of a low overpotential for hydrogen evolution at the iron electrode;
- frequent maintenance due to considerable gassing on charge.

For these reasons, the applications for nickel–iron batteries have been limited in scope. Many attempts to solve the gassing problem have not been successful. By contrast, the nickel–cadmium battery (see next section) has enjoyed greater success commercially as it is not subject to these limitations.

In the late 1970s and 1980s there was a revival of interest in nickel–iron as a traction battery for electric vehicles. The attraction lay in its higher specific energy than lead–acid and the belief that it could be commercialized quite quickly. Traction batteries were built and demonstrated by Eagle-Picher Industries in the USA, and by SAFT in France. Although the batteries performed as expected in road trials, when fitted to electric versions of the VW Golf and the Peugeot 205, their comparatively high

Table 9.2 *Specific energies of nickel–iron and lead–acid batteries at two discharge rates*

Discharge rate (W kg^{-1})	Nickel–iron (Wh kg^{-1})	Lead–acid (Wh kg^{-1})
20	54	36
40	50	26

cost, together with the high maintenance requirement, made them commercially unattractive.

9.4 NICKEL–CADMIUM BATTERIES

The nickel–cadmium battery employs the same nickel oxide positive electrode as the nickel–iron battery in combination with a metallic cadmium negative. Commercially, this has been a far more successful battery than nickel–iron and is extensively manufactured in all sizes from small consumer cells to large industrial batteries. Nickel–cadmium batteries are sometimes known colloquially as 'nicads'.

The cell charge–discharge reaction is analogous to that of nickel–iron, *viz.*

$$\text{Cd} + 2\text{NiOOH} + 2\text{H}_2\text{O} \underset{\text{Charge}}{\overset{\text{Discharge}}{\rightleftarrows}} \text{Cd(OH)}_2 + 2\text{Ni(OH)}_2 \qquad (9.2)$$

$$V^\circ = +1.30\,\text{V}$$

The practical, on-line voltage is 1.2 V. As the atomic mass of cadmium (112.4) is twice that of iron (55.8), the specific energy of the nickel–cadmium battery is not so high as that of nickel–iron. The practical specific energy of traditional nickel–cadmium batteries is 30 to 40 Wh kg^{-1}, *i.e.* rather similar to that of lead–acid batteries, although performance has been raised to 60 Wh kg^{-1} in some recent designs.

After lead–acid, nickel–cadmium is the most widely used rechargeable battery for industrial applications. Its high-rate and low-temperature performances are better than those of lead–acid. Other beneficial features of nickel–cadmium batteries are a flat discharge voltage, long life (~ 2000 cycles), continuous overcharge capability, low maintenance requirements, and excellent reliability. Cells and batteries are available in many different sizes and with pocket-plate, plastic-bonded, or sintered electrodes. A French company, Sorapec, has developed a nickel-felt electrode which is electrochemically impregnated with active material. Nickel–cadmium batteries using these electrodes show good charge-retention and charge-acceptance at high temperatures (60°C) where batteries of traditional design do not perform well. The principal disadvantages of the nickel–cadmium system are high cost (up to 10 times that of lead–acid) and environmental concerns associated with the disposal of batteries which contain toxic cadmium.

9.4.1   Industrial Batteries

Apart from the introduction of new types of positive electrode, the improvements associated with nickel–cadmium batteries over the past 50 years have been more in the area of cell design than in basic materials science. Traditional cells are of the vented variety and are sold in large sizes for various applications. With attributes of reliability and low maintenance, the battery is ideal for stand-by power duties, whilst the high power output, maintained at low temperatures, makes it suited to the starting of large engines where 'cold-cranking' currents of 5000 to 10 000 A may be required. Vented nickel–cadmium batteries also find applications on railways, on aircraft, for marine duties, and as power sources for mine locomotives and industrial trucks.

In France, particularly, there has been considerable interest in nickel–cadmium traction batteries for electric vehicles. For example, SAFT has opened a factory in Bordeaux to manufacture up to 10 000 vehicle traction batteries per year, both nickel–cadmium and nickel–metal-hydride. The batteries are being fitted to electric versions of the Peugeot 106 and the Renault Clio. In most respects, the performance of these vehicles meets the specifications for urban electric cars (see Section 12.4, Chapter 12). Typically, the daily range is 90 km, which is adequate for city use. It is argued that the high cost of nickel–cadmium batteries is mitigated by the much longer cycle-life compared with lead–acid and, moreover, the cost is likely to fall as the production rises. A 6 V (five cell) nickel–cadmium monobloc for use in electric vehicles is shown in Figure 9.3. The design incorporates sintered positive electrodes and plastic-bonded negative electrodes, and has a specific energy of 55 Wh kg^{-1} at the 3-h rate. The batteries are claimed to have an operational lifetime which exceeds 100 000 km in electric vehicles, and to be useable over a wide temperature range (-20 to $+40°C$).

The duties performed by nickel–cadmium batteries in some aircraft, both civil and military, range from starting the auxiliary power unit (at temperatures between -30 and $+70°C$) to powering all the on-board functions which include instruments and radio communication. In this application, reliability is essential. Varta Batterie AG manufactures single cells of 15 to 44 Ah capacity which are based on sintered positive and negative electrodes for maximum reliability. Up to 20 cells are series-connected in a single enclosure (Figure 9.4).

Figure 9.3 *Nickel–cadmium monobloc (6 V) for use in electric vehicles*
(By courtesy of SAFT)

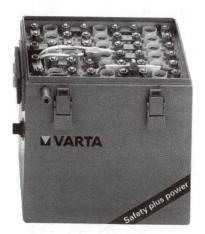

Figure 9.4 *Nickel–cadmium aircraft battery (24 V)*
(By courtesy of Varta Batterie AG)

9.4.2   Consumer Cells and Batteries

Small, sealed nickel–cadmium batteries have been used in electrical devices for over 40 years. The batteries can be used in any orientation and have capacities which range from 10 mAh to 10 Ah. Normally, three separators are used, namely, two wrappings of thin microporous separator, which is employed to hinder the growth of cadmium dendrites, and a thick 'reservoir separator' to hold the electrolyte (potassium hydroxide). The reservoir separator is often a cellulosic material which allows oxygen liberated on charge to pass freely to the negative electrode where it is reduced to water. The sealed cells are analogous to the valve-regulated type of lead–acid battery (see Section 8.7, Chapter 8) as regards their recombination technology and also in having safety valves to relieve excess pressure. They are manufactured as button cells (10 mAh to 2 Ah), cylindrical cells (100 mAh to 10 Ah), and prismatic cells. Cylindrical cells come in standard sizes and may be used as replacements for primary alkaline-manganese cells, though with inferior performance because of reduced capacity and lower voltage (Table 9.1). A five-cell, nickel–cadmium battery (6 V) replaces a four-cell, alkaline-manganese battery. The cylindrical cell may have either a spiral ('jellyroll') construction, for high power output (Figure 9.5), or a bobbin construction, for higher volumetric capacity. Sealed nickel–cadmium cells find their principal use in consumer applications, although special versions have been designed for telecommunications, and also for satellites and spacecraft.

Nickel–cadmium consumer cells may be recharged in 14 to 16 h after being completely discharged. A constant-current charger at the 10-h rate is suitable. When the cell reaches top-of-charge, any further passage of current will result in the evolution of oxygen at the positive electrode and hydrogen at the negative. These gases then recombine on the negative electrode with the liberation of considerable heat. Cells should not be recharged at too high a rate for this reason, and also because the rate of gassing will exceed the rate at which the gases can recombine and pressure will build up inside the cell until the vent opens.

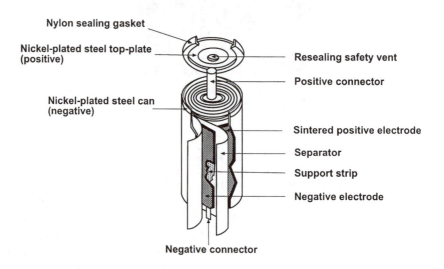

Figure 9.5 *Sealed, cylindrical, nickel–cadmium cell* (By courtesy of Ever Ready)

Sealed nickel–cadmium cells exhibit an unfortunate characteristic known as the 'memory effect'. When cells are only partially discharged, and then recharged, they progressively lose their capacity, cycle by cycle. To restore the lost capacity, it is necessary to 'condition' the cells by discharging completely to zero volts and then recharging fully. The memory effect is not observed in industrial vented cells, nor is it present in competing consumer cells such as rechargeable alkaline-manganese and nickel–metal-hydride.

Nickel–cadmium cells are the well-established choice for consumer applications where there are requirements for high current drain and long cycle-life, and where first cost is not a prime consideration. Typical applications are heavy duty motor-driven appliances, toys, and emergency systems. The cells have a flat discharge curve, which results in constant power throughout the discharge, and may be used over a wide temperature range, *i.e.* -20 to $+40°C$ (or even $+50°C$ with certain cell designs). Disadvantages are: lower voltage than primary cells, lower capacity (Table 9.1), slow overnight recharge, relatively high self-discharge rate, and the memory effect. Also, there is growing environmental concern over the toxicity of cadmium. In many countries, for example, spent cells can no longer be discarded with domestic waste. Progressively, these cells are likely to be replaced by rechargeable alkaline-manganese for low-rate applications, and by nickel–metal-hydride for high-rate applications. At the present time, nickel–metal-hydride and lithium-ion batteries have almost entirely taken over the premium markets of mobile telephones and portable computers, while nickel–cadmium is still holding its own in the lower cost markets of portable tools and toys.

9.5 NICKEL–ZINC BATTERIES

Zinc is the ideal material for negative electrodes in alkaline electrolyte batteries on account of its high electrode potential. It is the most electropositive of the common metals which can be plated from aqueous solution. The nickel–zinc cell has therefore a comparatively high voltage, *i.e.*

$$Zn + 2NiOOH + 4H_2O \underset{\text{Charge}}{\overset{\text{Discharge}}{\rightleftarrows}} Zn(OH)_2 + 2Ni(OH)_2 \cdot H_2O \qquad (9.3)$$

$$V° = +1.78\,V$$

and a correspondingly high specific energy. A practical nickel–zinc cell discharges at $\sim 1.6\,V$ and can attain 90 to 100 Wh kg^{-1}, while an industrial battery pack would be expected to yield 70 Wh kg^{-1}. This performance is substantially higher than that of lead–acid, nickel–iron or nickel–cadmium, and is extremely attractive for electric vehicles because it virtually doubles the daily range provided by lead–acid batteries. Moreover, zinc is considerably cheaper than cadmium and is non-toxic.

The nickel–zinc battery has been extensively studied in recent years as a candidate traction battery for electric vehicles. Unfortunately, the system suffers from one serious drawback: the greater solubility of zinc in potassium hydroxide, compared with cadmium or iron, leads to a much reduced cycle-life. This solubility causes the zinc to migrate and accumulate towards the centre of the negative plate which, thereby, changes shape and loses capacity. Also, there is a marked tendency for zinc dendrites (needles) to grow from the negative during recharging and these can penetrate the separator and cause an internal short-circuit. Much research has been directed towards elucidating and overcoming these limitations, and some success has been reported. Prototype batteries with stabilized zinc electrodes have achieved up to 500 cycles, but this performance is still too low for commercial success as a traction battery for electric road vehicles. If the problem of short life can be resolved, then it would appear that the nickel–zinc battery has a bright future. This is one of the remaining major challenges in the field of alkaline batteries. For the present, Yuasa in Japan produces the Yuni-Z range in 6.8 and 13.6 V monoblocs with a specific energy of ~ 60 Wh kg^{-1}. In the USA, Evercel Inc. is manufacturing nickel–zinc traction batteries for electric scooters and bicycles. The units are said to perform as well as nickel–metal-hydride batteries and are cheaper, but are not yet available for electric cars.

9.6 NICKEL–METAL-HYDRIDE BATTERIES

Nickel–metal-hydride cells are a later development of the nickel–hydrogen battery employed in satellites (see Section 12.2, Chapter 12). The hydrogen is stored reversibly in the form of a metal hydride which forms the negative electrode of the cell. The positive electrode is a standard nickel oxide electrode and the electrode reactions are as follows:

At the positive electrode:

$$\text{NiOOH} + \text{H}_2\text{O} + \text{e}^- \underset{\text{Charge}}{\overset{\text{Discharge}}{\rightleftarrows}} \text{Ni(OH)}_2 + \text{OH}^- \tag{9.4}$$

At the negative electrode:

$$\text{MH}_x + \text{OH}^- \underset{\text{Charge}}{\overset{\text{Discharge}}{\rightleftarrows}} \text{MH}_{x-1} + \text{H}_2\text{O} + \text{e}^- \tag{9.5}$$

To provide an effective storage medium for hydrogen, the hydride should have the following properties:

- high storage capacity;
- ready formation and decomposition at an appropriate rate;
- repeated cycling capability without change in pressure–temperature characteristics and with low hysteresis;
- good corrosion resistance;
- proven safety in use;
- low cost.

This is a formidable set of specifications, but it has now been met.

Early work, in the 1970s, utilized LaNi_5 alloy as the negative; this takes up hydrogen reversibly at ambient temperature to form LaNi_5H_6. The alloy has a dissociation pressure of ~ 100 kPa (1 atmosphere) at 15°C. Later commercial development work has focused on two new alloys to store hydrogen, as follows.

(i) A complex alloy based upon 'misch metal' (an unrefined mixture of rare earth elements), with various additives to adjust the dissociation pressure to the desired value and also to form a surface oxide film which acts as a barrier to prevent progressive oxidation of the metal hydride. This proprietary formulation is known as an AB_5-type alloy, where A is a mixture of rare earths and B is partially-sub-stituted nickel.

(ii) A similarly complex, multi-component alloy of the type AB_2, where A is titanium and/or zirconium and B is again partially-substituted nickel.

The AB_2 alloy is claimed to have a higher capacity for hydrogen storage, to provide superior oxidation and corrosion resistance, and to be less costly.

The operating voltage of a nickel–metal-hydride cell is almost the same as that of nickel–cadmium (1.2 to 1.3 V), which allows ready interchange-ability, and the discharge curve is quite flat. The capacity of a nickel–metal-hydride cell is significantly greater than that of a nickel–cadmium cell of the same size (Figure 9.6), with the result that the specific energy (60 to 70 Wh kg^{-1}) is 1.5 to 2 times higher. Moreover, the specific power of

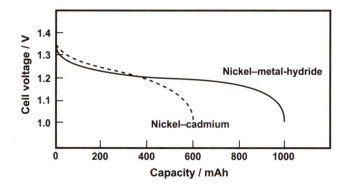

Figure 9.6 *Discharge curves for AA-size nickel–metal-hydride and nickel–cadmium cells*
(5-h rate)
(By courtesy of Research Studies Press Ltd)

nickel–metal-hydride batteries may be as high as 250 W kg^{-1}. The batteries are resilient to both overcharge and overdischarge, and may be operated from -30 to $+45°C$. Another attraction is that there are no toxicity problems associated with recycling. Disadvantages are a comparatively high cost, a higher self-discharge rate than nickel–cadmium, and poor charge-acceptance at elevated temperatures.

Cells of both the cylindrical (spirally-wound) and the prismatic design are now manufactured in a range of sizes. Small cells are used in portable electronic devices (*e.g.* mobile telephones, laptop computers) and in portable tools, while prismatic cells of ~ 100 Ah are available for assembly into 12 or 24 V modules (*e.g.* for use as traction batteries). Coin cells are also produced by most of the Japanese manufacturers. The new materials technology involved in the development of nickel–metal-hydride batteries is almost entirely associated with the hydride negative electrode, and it is said to be comparatively easy to switch a production line from nickel–cadmium cells to nickel–metal-hydride cells.

By the end of 1997, the installed production capacity in Japan alone was 50 million cells per month, which were mostly small units for electronics applications. Annual world-wide sales increased by 33% between 1997 and 1998 to US$900 million. This growth has arisen through the entry of many battery companies in Europe and the USA into the market. For example, Duracell manufactures small batteries of voltage 3.6, 4.8, 6.0 and 10.8 V (3, 4, 5 and 9 cells, respectively) and of capacities from 1.5 to 3.0 Ah. In France, SAFT is making large 12 and 24 V monoblocs for electric-vehicle traction applications (Figure 9.7). These monoblocs have a capacity of 93 Ah and a specific energy of 64 Wh kg^{-1} (3-h rate). A nickel–metal-hydride battery pack has been designed specially for the Chrysler

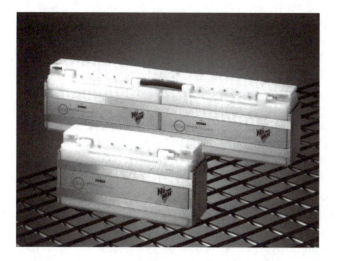

Figure 9.7 *Nickel–metal-hydride batteries for electric-vehicle traction. (Foreground: 12 V unit. Background: 24 V unit)*
(By courtesy of SAFT)

'Epic' electric minivan and provides a daily range of 150 km. The pack is water-cooled to keep the operating temperature below 50 °C. In Germany, Varta Batterie AG has made traction batteries based either on high-energy modules (75 Wh kg^{-1}, 160 W kg^{-1}) or high-power modules (55 Wh kg^{-1}, 300 W kg^{-1}). The batteries have been installed in a number of different electric and hybrid electric cars. In Japan, nickel–metal-hydride batteries produced by Panasonic EV Energy have been fitted to an electric vehicle (Toyota 'RAV-EV') and to hybrid electric vehicles (Toyota 'Prius' and Honda 'Insight').

The rate at which the nickel–metal-hydride battery has penetrated the market in recent years is quite phenomenal. The technology has taken over both new applications and part of the share formerly occupied by nickel–cadmium. Looking to the future, it appears likely that nickel–metal-hydride will remain one of the world's major rechargeable batteries, although there will be a strong challenge from lithium-ion batteries (see Chapter 10) which also are making remarkable progress.

9.7 SILVER OXIDE BATTERIES

Silver batteries contain Ag(II) oxide as the positive electrode. The discharge proceeds in two steps, namely, $Ag^{2+} \rightarrow Ag^{+}$ and $Ag^{+} \rightarrow Ag$, at voltages of 1.85 and 1.59 V, respectively. This feature presents the possibility of batteries with higher specific energy than those with a positive electrode of nickel oxide which discharges only in a one-electron step to nickel hydroxide. Of the four possible silver oxide batteries, silver–cadmium, silver–hydrogen and silver–iron have received relatively little attention and have not assumed great commercial significance. On the other hand, rechargeable silver–zinc batteries have found application where high levels of specific energy and specific power are critical, notably in the military sphere. Examples are the supply of propulsion power to torpedoes and submersibles.

In addition to the attributes of high specific energy and high energy density, *viz.* ~ 150 Wh kg^{-1} and 200 Wh dm^{-3}, the silver–zinc battery can be operated over a wide range of temperature (-20 to $+70°C$). Useful life is limited, however, to a few tens of cycles at best, on account of the significant solubility of both ZnO and Ag_2O_2 in potassium hydroxide. This fact, together with the high cost of silver, severely restricts the applications for the battery. Nevertheless, prismatic cells and batteries are manufactured by a number of firms, *e.g.* Eagle Picher (USA), SAFT (France), and Yuasa (Japan).

An interesting demonstration of the capabilities of the silver–zinc battery has been in the World Solar Challenge. This is a race for lightweight, solar-powered electric cars across the 'Outback' of Australia from Darwin to Adelaide (3000 km). Each car is permitted to have up to 5 kWh of batteries to store electricity obtained from its solar cells. The winners of all the five races held to date have been powered by silver–zinc battery packs.

Chapter 10

Lithium Batteries

10.1 INTRODUCTION

The general principles which govern lithium batteries have been outlined in Chapter 6 when discussing primary versions of these batteries. In summary, lithium is extremely attractive as a negative electrode material on account of its low atomic mass and high electrode potential. Because of its reactivity towards moisture, it is necessary to handle the metal and to construct cells in a dry-room, and also to employ a non-aqueous (organic) liquid electrolyte with a dissolved lithium salt to provide adequate ionic conductivity. A wide variety of organic liquids and dissolved salts has been investigated. The positive active-material in primary lithium batteries is generally an inorganic oxide or sulfide, or sometimes a polycarbon fluoride (CF_x) made by immobilizing fluorine in a graphite host.

The rechargeable lithium cell has proved to be much more difficult to develop than the primary cell. It is only since 1990 that practical cells of the 'lithium-ion' type (see Section 10.4), with graphite negative electrodes, have become commercially available. Before that, throughout the 1970s and 1980s, research efforts focused on developing a rechargeable version of the primary cell that used lithium-metal foil for the negative electrode. Some success was achieved with this venture, though not sufficient to justify large-scale manufacture.

10.2 INTERCALATION ELECTRODES AND LITHIUM-METAL CELLS

Research carried out at the Exxon Laboratories (USA) in the early 1970s showed that lithium ions are capable of being electrochemically 'intercalated' into the crystal lattice of certain inorganic compounds. The classic example is titanium disulfide (TiS_2) and the reaction may be represented by:

$$xLi^+ + TiS_2 + xe^- \underset{\text{Charge}}{\overset{\text{Discharge}}{\rightleftarrows}} Li_xTiS_2 \quad \text{where: } 0 < x < 1 \qquad (10.1)$$

Titanium disulfide has a 'layered' crystal structure and the lithium ions insert themselves between the layers. The reaction takes place when titanium disulfide is made the positive electrode in a cell which uses a solution of a lithium salt as the electrolyte. The intercalation of lithium is rapid and fully reversible. There is no recrystallization or phase change, as usually occurs when a positive electrode is discharged; the lithium ions are simply taken into the lattice structure of the titanium disulfide with only a small change in lattice parameter. This provides the basis for a positive electrode which can be discharged (lithium ions inserted) and recharged (lithium ions removed) at rates which give currents in excess of $10\ mA\ cm^{-2}$. The essential features of titanium disulfide for this application are: a high capacity for lithium, rapid intercalation and de-intercalation, no significant change in crystal structure, good electrical conductivity. One consequence of the fact that the crystal structure remains intact, and only a single solid phase is involved, is that thermodynamics dictate a declining voltage curve as discharge proceeds. This is more akin to the behaviour of a primary zinc–carbon cell rather than to that of a secondary nickel–cadmium cell.

Stemming from this original discovery for titanium disulfide, there has been much research on intercalation compounds, particularly for lithium ions in the context of rechargeable lithium batteries. Other promising compounds are a vanadium oxide (V_6O_{13}), molybdenum disulfide (MoS_2), and molybdenum trisulfide (MoS_3). Researchers at the AT&T Bell Laboratories (USA) focused on niobium triselenide ($NbSe_3$) as a positive electrode. This material undergoes the following reversible reaction:

$$3Li^+ + NbSe_3 + 3e^- \underset{\text{Charge}}{\overset{\text{Discharge}}{\rightleftarrows}} Li_3NbSe_3 \qquad (10.2)$$

Up to three lithium ions are incorporated per formula unit of $NbSe_3$. Using a negative electrode made from lithium foil, this cell has an open-circuit voltage of 2.2 V and may be discharged down to 1.4 V. Compared with a nickel–cadmium cell of similar size, it has almost twice the capacity as well as a higher voltage, so that its stored energy content (300 Wh dm^{-3}) is two to three times greater (Figure 10.1). AT&T named the cell the 'Faraday^{TM} cell' and set up a pilot line for the production of AA-size units. Cycle-lives in excess of 200 cycles were demonstrated and the self-discharge rate was very low, with a shelf-life of more than five years. Despite this encouraging performance, and the investment in the pilot manufacturing facility, the project was ultimately abandoned.

The first rechargeable lithium cell to be commercialized, in the late 1980s, was manufactured in Canada by the Moli Energy Corporation of British Columbia. This was based on the Li–MoS$_2$ electrochemical couple and was known as the Molicel^{TM}. It was a spirally-wound cell of lithium foil, a separator and the mineral molybdenite (naturally occurring MoS$_2$) as the positive electrode, with the usual electrolyte of a lithium salt dissolved in a polar organic liquid (*i.e.* one with unbalanced electric charges to give a significant dipole moment). The cell had a nominal voltage of 1.8 V and a specific energy which was two to three times higher than either lead–acid or nickel–cadmium. Service-lives of 200 to 300 cycles were demonstrated. The cell was manufactured in both AA- and C-sizes, and some larger cells (65 Ah) were constructed experimentally. Ultimately, the battery was withdrawn from the market after safety difficulties were experienced with overheating on recharge.

10.3 REPLATING OF LITHIUM

The recharge reaction at the negative electrode of a secondary battery is normally a matter of reducing a metallic salt to the free metal. If the salt is essentially insoluble, then no particular problems arise. In the lead–acid battery, for example, the reduction process takes place within the individ-

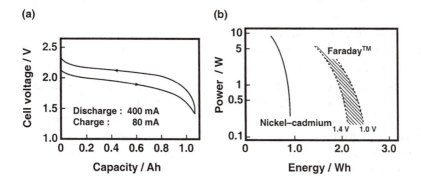

Figure 10.1 (a) *Discharge curve and* (b) *electrical performance characteristics of AA-size, Li–NbSe$_3$ cells (the Faraday^{TM} cell) at 20°C*
(By courtesy of AT&T Corp.)

ual grains of the discharge product (lead sulfate) attached to the negative electrode and this is a solid-state reaction. If the salt is appreciably soluble, however, then the reduction process becomes a question of electroplating a metal from solution onto the negative electrode. It is here that difficulties can arise, as have been discussed earlier for the zinc electrode in the nickel–zinc battery (see Section 9.5, Chapter 9). The plated metal can redistribute itself across the electrode in a non-uniform manner and also grow dendrites which lead to internal short-circuits. With lithium cells, the situation is far worse than for zinc, as *all* the discharged lithium is accommodated in the positive electrode, and has to re-cross the electrolyte as lithium ions during recharge and be electroplated out once again on the negative electrode. This increases the possibility of non-uniform distribution of the metal across the electrode together with the development of dendrites.

A further complication with lithium cells is that the metal is thermodynamically unstable with respect to the electrolyte and is therefore liable to react with it. As discussed in Section 6.2, Chapter 6, the long shelf-life of primary lithium batteries is attributable to a thin passivating layer of hydroxide–nitride, formed on the surface of the lithium foil, which acts as a secondary solid electrolyte and protects the metal from reacting with the organic electrolyte. With a secondary battery, this situation prevails up to and during the first discharge, but, immediately recharge begins, finely-divided lithium is deposited on the negative electrode. The smoothness, or otherwise, of the electroplate is greatly influenced by the nature of the electrolyte and the plating conditions. The lithium deposit is unstable with respect to the electrolyte and will tend to react with it to a greater or lesser degree. This reaction can result in undesirable consequences. At best, it will lead to capacity loss as any lithium which reacts will no longer be available for a subsequent discharge.

A more serious situation develops when clumps of re-plated lithium microcrystals become electrically isolated from the electrode and so, again, are not available for discharge. The worst scenario of all is when the reaction with the electrolyte is so rapid that the associated heat generation results in thermal runaway, melting of the lithium, and a fire in the cell. The same sequence of events may also arise as a result of an internal short-circuit caused by penetration of the separator by dendrites. A number of secondary lithium cells with metal-foil electrodes have caught fire during recharge and this may be one reason why they are not commercially available today. Evidently, the electroplating of lithium metal and the recharging of lithium batteries is not a straightforward matter and is one which is still subject to on-going research.

10.4   THE LITHIUM-ION BATTERY

The origin of the lithium-ion battery lies in the discovery, made by researchers at Oxford University in the UK in the late 1970s, that lithium ions can be intercalated (absorbed) into the crystal lattice of trivalent cobalt or nickel oxides to give the compounds $LiCoO_2$ and $LiNiO_2$, respectively. When these oxide electrodes were used as positives against metallic lithium negatives in an organic electrolyte, a 4 V cell resulted. This discovery was followed up by workers at the Sony Corporation in Japan who realized that with such a high cell voltage it would be possible to employ two intercalation electrodes – one positive and one negative – with lithium ions shuttling back and forth between the two. This cell, the 'lithium-ion cell', contains no metallic lithium and is therefore much safer on recharge than the earlier, lithium-metal design of cell. The use of an intercalation compound as the negative electrode will inevitably result in a cell voltage which is lower by an amount that corresponds to the free energy of dissolution of lithium in the electrode. Nevertheless, one could afford to lose up to a volt at the negative electrode and still have a very respectable 3 V cell. This opened up the possibility of a practical 3 V lithium-ion cell. In effect, there is a trade-off between a lower cell voltage and a far simpler system with no lithium metal present.

It was soon found that the preferred host for the negative electrode was carbon, in the form of either graphite or an amorphous material with a high surface-area such as coke. The intercalation of chemicals between the carbon planes of graphite had long been known. Experiments showed that graphite was capable of hosting lithium up to a composition of LiC_6. The voltage, with respect to a lithium reference electrode, varied from zero for the fully intercalated LiC_6 to about $+1$ V for graphite from which all the lithium has been removed. The use of carbon was ideal on two counts. First, carbon is a readily available and cheap material of low mass. Second, carbon takes up a respectable quantity of lithium with a voltage in just the right range so that, when paired with a transition metal oxide as the positive electrode, it gives a cell with a voltage which starts at 4 V in the fully charged state and declines to 3 V during discharge. The electrode reaction may be represented by:

$$Li_xC_6 \underset{\text{Charge}}{\overset{\text{Discharge}}{\rightleftharpoons}} 6C + xLi^+ + xe^- \quad \text{where: } 0 < x < 1 \qquad (10.3)$$

Carbon can be obtained from many different sources, *e.g.* graphite, petroleum coke, carbon blacks, pyrolysed polymers. The different sources of carbon vary in their ability to intercalate lithium ions. Early work established that carbons from petroleum coke were only able to take up half as much lithium as graphite (corresponding to $Li_{0.5}C_6$). In the first intercalation cycle of graphite, the measured capacity is greater than that corresponding to LiC_6, and greater than that obtained in subsequent cycles.

During intercalation, the electrolyte undergoes partial decomposition which is associated with expansion and exfoliation of the graphite. The decomposition products are adsorbed on the graphite to form a passivating surface layer which inhibits further decomposition. Coke electrodes, although having lower capacity than graphite, have the advantage of being less sensitive to the nature of the electrolyte employed and permit the use of electrolytes of higher conductivity. Nevertheless, most lithium-ion cells presently employ graphite electrodes as they give a flatter discharge curve. Lightweight metal alloys are being investigated as possible alternatives to carbon for use as negative intercalation electrodes.

The concept of the lithium-ion cell is illustrated schematically in Figure 10.2. The lithium ions 'swing' or 'rock' backwards and forwards between one electrode and the other as the battery charges and discharges. Accordingly, these cells have also been known as 'swing cells' and 'rocking-chair cells'. Now, however, 'lithium-ion cell' is the preferred name.

The majority of commercial lithium-ion cells employ positive elec-

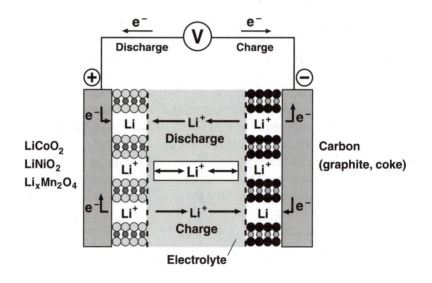

Figure 10.2 *Schematic representation of the charge–discharge operation of a lithium-ion cell*
(By courtesy of Research Studies Press Ltd.)

trodes of cobalt oxide. These prove to be the most satisfactory technically, although also the most expensive. The positive electrode reaction is:

$$Li_{0.55}CoO_2 + 0.45Li^+ + 0.45e^- \underset{\text{Charge}}{\overset{\text{Discharge}}{\rightleftarrows}} LiCoO_2 \qquad (10.4)$$

where $0.45Li^+$ is the maximum amount of lithium ions that can be de-intercalated from $LiCoO_2$. The cheaper positive electrode based on nickel oxide, used by some manufacturers, has a somewhat more complex intercalation–de-intercalation process which involves subtle structural rearrangements. Nevertheless, it behaves satisfactorily. In this case, the positive electrode reaction is:

$$Li_{0.35}NiO_2 + 0.5Li^+ + 0.5e^- \underset{\text{Charge}}{\overset{\text{Discharge}}{\rightleftarrows}} Li_{0.85}NiO_2 \qquad (10.5)$$

When either of the above positive electrodes is combined with a LiC_6 negative electrode, a cell with a nominal voltage of 3.6 V is obtained. Thus, in terms of voltage alone, one lithium-ion cell is equivalent to three nickel–cadmium or three nickel–metal-hydride cells.

Other possible positive electrodes are based on manganese oxide, namely, $LiMnO_2$ and $LiMn_2O_4$. The latter compound has a spinel structure $Li[Mn_2]O_4$. When lithium ions are de-intercalated from it, corresponding to the oxidation of Mn^{3+} to Mn^{4+}, a 4 V cell is obtained *versus* lithium metal. When, however, lithium ions are inserted into $LiMn_2O_4$, moving in the direction of $LiMnO_2$, a 3 V cell results. This transfer of lithium into and out of the structure causes subtle and complex shifts in the detailed crystallographic structure. It is therefore difficult to prepare the manganese compounds in reproducible form; a situation which is considerably more complex than with $LiCoO_2$ or $LiNiO_2$. On the other hand, manganese is widely available, is much cheaper than nickel or cobalt, and is non-toxic – three good reasons why it is used in consumer primary batteries. Accordingly, much research is being directed towards developing a fully satisfactory, manganese-based, positive electrode for use in lithium-ion cells.

Lithium-ion batteries were first introduced commercially in 1991 by the Sony Corporation in Japan. Other Japanese manufacturers soon entered the market, followed closely by American and European companies. The subsequent growth in sales of the batteries was truly phenomenal. By 1998, in Japan alone, 190 million lithium-ion cells valued at

US$2.1 billion were being manufactured annually. These cells were principally for use in mobile telephones, portable computers and camcorders, as well as in other electronic circuits. Today, lithium-ion cells are being manufactured in many different countries. World-wide production is estimated at 500 million units per annum.

Why is lithium-ion proving to be so popular? In part, it is the enormous expansion in the market for portable electronic devices, and in part the desirable characteristics of the battery. The attributes of lithium-ion may be summarized as follows:

- high energy: gravimetric energy density ~ 125 Wh kg^{-1}; volumetric energy density ~ 300 Wh dm^{-3};
- high average operating voltage (3.6 V);
- excellent charge–discharge characteristics, with more than 500 cycles possible;
- acceptably low self-discharge on standing ($< 10\%$ per month);
- absence of a memory effect as found with nickel–cadmium batteries (see Section 9.4.2, Chapter 9);
- easy determination of remaining capacity by virtue of having a sloping discharge curve;
- much safer than equivalent cells which use lithium metal, with no special transportation regulations;
- rapid recharging is possible, *i.e.* within just 2 to 3 h.

It is these features which have made the lithium-ion battery so popular in recent times.

As with all batteries, there are some undesirable aspects. For example, the replacement of lithium metal by C_6Li in the negative electrode has led to a substantial loss of specific energy, in return for a cell with much longer cycle-life and which is safer in use. Even so, extreme care has to be exercised in controlling the charging conditions, especially the top-of-charge voltage – the limit must not exceed 4.2 V for $LiCoO_2$ or 4.1 V for $LiNiO_2$. A further concern is that overcharging or heating above $\sim 100\,°C$ results in decomposition of the charged positive electrodes (low values of Li_x) with liberation of oxygen gas. Li_xNiO_2 leaves $LiNi_2O_4$ (a spinel), while Li_xCoO_2 yields Co_3O_4. Not only does electrolyte decomposition ruin the cell, but also the liberation of oxygen gas could be dangerous, particularly if the cell vent failed to operate correctly.

In the manufacture of lithium-ion cells it is necessary to employ an electrolyte (organic liquid plus dissolved salt) which is stable towards oxidation at 4.2 V. The positive and negative active-materials are applied uniformly to both sides of thin metal foils (aluminium and copper, respectively). Binders are used as adhesives. Cell assembly involves spirally winding the electrode sheets with a microporous polymer sheet between them to act as the separator. The operation has to be conducted in a dry-room to ensure that the cell is completely free of water. In the top of the cell is a safety vent, set to open at a predetermined pressure, and a safety element (with a positive temperature coefficient of resistance, see Section 6.3, Chapter 6) which cuts off the charging current if it becomes too large. The cells are assembled in the discharged state and have to be charged before being put into service.

Charging of lithium-ion cells is carried out at constant current until the top-of-charge voltage is reached at about 80% of the charge capacity. Thereafter, charging continues at constant voltage until the current declines to a low value. The cell is fully charged after 2 to 3 h. Charging should be performed in the temperature range 0 to 40°C. Particular care is needed when lithium-ion cells are used in series or parallel combinations in a battery, since overcharge or cell reversal may cause safety problems. Battery protection circuits, as recommended by the manufacturers, are normally built into the battery pack. Typically, a field-effect transistor opens if the charge voltage of any cell reaches 4.30 V, and a fuse activates if the cell temperature becomes too high. If a safe pressure threshold is exceeded, a pressure switch interrupts the charge current. Most manufacturers do not sell lithium-ion cells as individual units, but make them available in a battery pack, complete with a protection circuit.

Lithium-ion cells are produced in coin format, as well as in cylindrical and prismatic shapes. The cylindrical cells are made in standard AA-size, of nominal capacity 580 mAh, and in larger non-standard sizes for professional use. The latter have capacities of 750, 900 and 1350 mAh. Discharge curves for the 1350 mAh cell manufactured by the Sanyo Corporation are shown in Figure 10.3. The curve at the $1C$ rate (the 1-h rate) has a declining voltage from 4 V initially to 3 V at the end of discharge, with a mean value of ~ 3.6 V. At the $2C$ (30-min) rate, the mean voltage drops to ~ 3.3 V. These, and most other, lithium-ion cells are capable of 500 charge–discharge cycles with loss of 20 to 30% capacity. Lithium-ion cells are also operable over a wide temperature range. The discharge curves ($0.2C$ rate) for the Sanyo cell between -20 and $+60$°C are given in Figure 10.4.

Other manufacturers are making even larger cells. For example, AGM Batteries (a joint UK–Japanese company) manufactures 4.5 Ah (D-size) cells. SAFT makes 6.5 and 13 Ah 'high-power' cells, as well as 44 Ah 'high-energy' cells. The characteristics of the SAFT cells are summarized in Table 10.1. The cells have nickel oxide ($LiNiO_2$) in the positive electrodes. The recommended temperature range of operation is -10 to

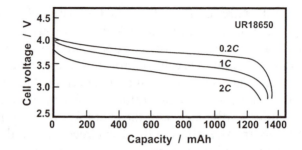

Figure 10.3 *Discharge curves of a 1350 mAh Sanyo lithium-ion cell at 25°C. (Note: this is a non-standard cell size; the code number UR18650 indicates that the cell is 18 mm in diameter and 65 mm in length)*
[By courtesy of Sanyo Energy (UK) Co. Ltd]

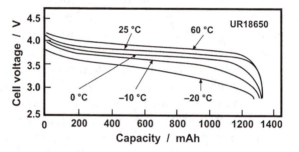

Figure 10.4 *Discharge curves for a 1350 mAh Sanyo lithium-ion cell as a function of temperature. Discharge at 0.2C rate*
[By courtesy of Sanyo Energy (UK) Co. Ltd]

Table 10.1 *Characteristics of SAFT lithium-ion cells*

	High-power cells		High-energy cell
Capacity at $C_3/3$ rate (Ah)	6.5	13	44
Specific energy (Wh kg^{-1})	64	70	144
Energy density (Wh dm^{-3})	135	150	308
Specific power (W kg^{-1})	1500	1350	300
Power density (W dm^{-3})	3100	2900	642

+45°C, and the voltage limits are 3.9 V on charge and 2.1 V on discharge.

For use as electric-vehicle traction batteries, SAFT manufactures 950 Wh modules which contain six of the 44 Ah cells arranged in any of three different series–parallel configurations. Even larger modules for electric vehicles have been made by the Sony Corporation. For example, a 2.9 kWh system, composed of eight 100 Ah cells with LiCoO$_2$ positive electrodes, has given a specific energy of 100 Wh kg^{-1}, an energy density of 160 Wh dm^{-3}, a specific power of 300 W kg^{-1}, and a life of 1200 cycles. Clearly, these are characteristics which make the unit exceptionally well suited as a power source for electric vehicles.

Particular care is needed when using large cells in battery packs. Since lithium-ion cells do not have a shuttle mechanism to prevent overcharge, cell imbalances within a series string can lead to overcharge which, in turn, results in corrosion, reduced cell life, and safety incidents. Similarly, cell imbalances can give rise to overdischarge which further aggravates the imbalance and causes more overcharge on the next charging cycle. For this reason, the voltage cut-off protection and electronic control circuitry described above are essential with packs of large cells.

To summarize, lithium-ion cells and batteries have made impressive technical progress within the short space of a decade, and even more remarkable commercial progress in terms of manufacturing throughput and market penetration. There seems little doubt that this is a battery for the 21st century, and one whose prospects will be enhanced still further if a satisfactory positive electrode based on the cheaper manganese oxide can be successfully developed.

10.5   THE LITHIUM–POLYMER BATTERY

The science of solid-state electrolytes, that is solids which exhibit a high conductivity for cations or anions but are electronic insulators, developed rapidly in the 1970s. As part of the programme of academic research, workers at the University of Grenoble in France investigated ionic conduction in polymers. It was found that polymers containing a heteroatom, such as oxygen or sulfur, tended to have a relatively high dielectric constant and to dissolve lithium salts in reasonably high concentrations. Subsequent work focused on poly(ethylene oxide), PEO, which dissolves salts such as lithium perchlorate (LiClO$_4$) and lithium trifluoromethane sulfonate (LiCF$_3$SO$_3$). The conductivity of these solid solutions for lithium ions is too low at ambient temperature, *viz.* $\sim 10^{-5}$ S m^{-1}, for the solutions to be of any use as battery electrolytes. On warming above about 60°C, however, there is a transformation of crystalline regions in the polymer, which essentially do not conduct ions, to amorphous regions which display much better electrical conductivity. In fact, the conductivity is improved by several orders of magnitude and reaches a value of $\sim 10^{-1}$ S m^{-1} at 100°C. This value is still low compared with those for conventional aqueous electrolytes, but does just about allow the polymer to serve as an electrolyte for lithium batteries, provided it is sufficiently thin. Calculations show that for an acceptable voltage drop of 10 mV across the electrolyte, the thickness of the polymer would need to lie in the range 10 to 100 μm.

The above studies gave rise, in the mid-1970s, to the concept of the lithium–polymer battery – a quite revolutionary concept in terms of battery design. The aim was to have an ultra-thin battery, 100 to 200 μm thick, in the form of a laminate of a lithium-metal foil (the negative electrode), a thin sheet of polymer electrolyte, and an intercalation compound (*i.e.* a metal oxide) deposited as a thin layer on a metallic current-collector (the positive electrode). For the system to function as a solid-state battery it is necessary that the lithium ions which exit the polymer electrolyte be brought into intimate contact with the metal oxide particles. This can be achieved by fabricating the positive electrode as a composite material in which the active oxide is mixed with graphite (to provide electronic conduction) and dispersed in more of the solid-polymer electrolyte. In this way, the lithium ions are conducted through the electrolyte to reach the surface of the individual particles of active oxide.

Much research and development work on this conceptual battery was carried out in the 1980s at the Harwell Laboratory in the UK. The preferred polymeric electrolyte was a lithium salt dissolved in PEO, and the preferred intercalation electrode was a vanadium oxide (V_6O_{13}). The electrolyte was made by dissolving the PEO and the chosen lithium salt in a solvent (*e.g.* acetonitrile, CH_3CN) and producing a thin film by solution casting. The same procedure was adopted for the composite positive electrode, with the addition of insoluble V_6O_{13} and carbon to form a slurry which was solution cast onto a current-collector of nickel foil. Typically, the thickness of the polymer electrolyte and the composite positive electrode was 25 and 50 μm, respectively. The thickness of the lithium foil was determined by its availability and was generally present in excess. The three components of the cell were handled and laminated in a dry-room.

As the conductivity of the polymer electrolyte is low, even at 100 °C, the current density is restricted to < 1 mA cm^{-2} in order that the voltage drop across the cell should not be too great. This low current density is unimportant since it is compensated by the large surface area of the cell, which may be of the order of many square metres, and thus the cell may be discharged at a high rate. Once the laminate has been made, it may be configured in many different ways (Figure 10.5). Small, flat-plate cells were packaged in composite aluminium–plastic envelopes of the type used to hold powdered foodstuffs. The current tabs protruded through the hermetic seal. Cells of larger area could be rolled into cylinders, or folded as shown in Figure 10.5. When using the jellyroll configuration, an additional insulator layer is needed to isolate the lithium negative electrode from the current-collector of the positive electrode. A bipolar

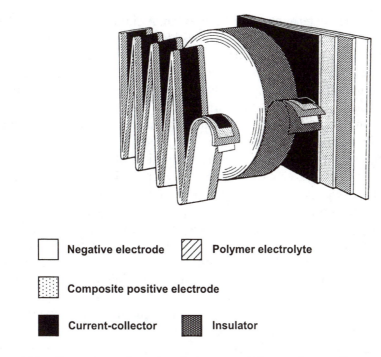

Figure 10.5 *Alternative configurations for large area, lithium–polymer cells*
(By courtesy of AEA Technology Batteries)

battery is made by backing the lithium negative electrode of one cell onto the current-collector of the positive electrode in the next cell, and so on. Cells of this type behaved well initially, but tended to lose their capacity fairly rapidly on cycling. The discharge curve showed a marked slope, *i.e.* the voltage decreased from the initial open-circuit value of 3.3 V to 1.7 V at the end of discharge. The cycle-life could be improved by operating over a narrower voltage range. The cells had some interesting characteristics. For instance, with care, it was possible to cut a flat cell into two with scissors so that each of the resultant pieces also continued to function as a cell. The thin lithium–polymer cell is seen as attractive for novel battery applications on account of the many ways in which it may be configured to suit the space available.

The principal disadvantage of the simple cell is the need to operate in the temperature range 80 to 120 °C. This seriously limits its applications. On the other hand, there are situations where high ambient temperatures exist (particularly in enclosed outdoor containers in summer) and where conventional aqueous batteries are short-lived. In such situations, a medium-temperature battery such as the lithium–polymer system should be ideal, as it may be easily thermally insulated and does not require the input of much external heat to keep warm. A similar argument has been advanced for the development of the lithium–polymer battery as a traction battery for electric vehicles. The temperature in vehicles varies enormously with geography and with season, and it is difficult to envisage an aqueous traction battery which will cope with all extremes of climate. By contrast, a battery which operates at around 100°C will be independent of the external temperature and, therefore, will not be difficult to maintain at operating temperature.

In recent times, the temperature restriction of the lithium–polymer battery has been largely overcome by incorporating a conventional organic liquid electrolyte into the polymer matrix to make an amorphous, so-called 'gelionic' electrolyte. When optimized, this electrolyte has most of the mechanical properties of a solid-polymer film and is essentially dry, but conducts lithium ions sufficiently well that a cell constructed from it may be used at ambient temperature. 'Gelionics', which may be regarded as immobilized liquid electrolytes, are often based on the polymer poly(vinylidene fluoride), PVDF.

During the 1990s, development and scale-up work on the lithium–polymer battery has continued at AEA Technology Batteries in the UK, and in North America through a collaborative effort between the 3M Company and the Argonne National Laboratory in the USA and Hydro-Quebec in Canada. Much of the scale-up work employs technology adapted from the coated paper industry, and involves a series of coating, bonding and lamination processes. Equipment, constructed in a dry-room, for preparing and coating electrolyte films is shown in Figure 10.6. It is amenable to high speed, high quality, automated manufacturing.

The development programme in North America has been supported by the United States Advanced Battery Consortium (USABC), a government–industry partnership, and is directed towards the production of traction batteries for electric vehicles. The chosen electrolyte consists of a lithium salt, $Li(CF_3SO_2)_2N$, dissolved in PEO and produces a polymer which, at 60 to 80°C, is able to achieve fast transport of lithium ions. The resultant battery is capable of giving a pulse power in excess of 200 W kg^{-1} for 30 s, as needed for vehicle acceleration. One of the attractions of the lithium–polymer technology is that by using the same laminate, produced in bulk, there is very great flexibility in the size and shape of the cells and battery modules that can be produced. For electric-vehicle applications, the team has produced a 119 Ah, 20 V module which weighs 15.7 kg (Figure 10.7). The module has a specific energy of 155 Wh kg^{-1}, an energy density of 220 Wh dm^{-3}, and a specific power of 315 W kg^{-1}. It is said to have a life in excess of 600 cycles and to be safe under all

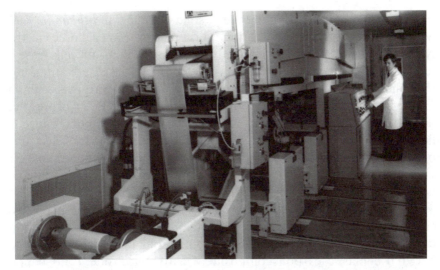

Figure 10.6 *Scale-up equipment for the quantity production of polymer and composite films (AEA Technology)*
(By courtesy of AEA Technology Batteries)

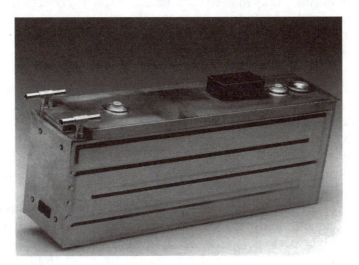

Figure 10.7 *Prototype lithium–polymer battery module (119 Ah) for electric-vehicle applications*
(By courtesy of the 3M–Argonne–Hydro-Quebec Consortium)

conditions of abuse. These reports lead to the conclusion that, by using a solid-polymer electrolyte, the USABC has succeeded in overcoming the recharging problems encountered in lithium-metal cells based on liquid electrolytes.

Other developers have taken the easier, and more commercially attractive, option of targeting the use of solid polymers in *lithium-ion* cells which operate at room temperature. This involves using gelionic electrolytes. These cells have most of the advantages of lithium-ion cells (ease of rechargeability, *etc.*) while retaining the configurational advantages of thin-film, laminate cells. Also, operation at room temperature is essential for use in portable electronic products. In the USA, the Bell Corporation has developed a plastic lithium-ion battery which can be bent, folded or flexed into any shape or configuration. It allows the designer of electronic products (camcorders, laptop computers, mobile telephones, pagers, video games, *etc.*) a new degree of freedom in how to accommodate the necessary battery. The laminate structure on which the cell is based is <0.5 mm thick and consists of five layers: (i) an aluminium mesh, positive current-collector; (ii) a plastic composite, positive electrode which contains lithium manganese oxide as active material; (iii) a gelionic electrolyte; (iv) a plastic–carbon composite, negative electrode; (v) a copper mesh, negative current-collector. The cell operates at 3.7 V (average) and retains 80% of its capacity after 1200 full discharge–charge cycles.

One of the first companies to commercialize solid-polymer lithium-ion batteries was Ultralife Batteries, Inc., which now manufactures a range of cells with capacities from 350 to 1700 mAh ($C_5/5$ rate). The cells are packaged in laminated foil jackets and, after packaging, their overall thickness varies from 3.2 to 6 mm, although cells as thin as 1 mm are possible. As an example, the 800 mAh cell is prismatic with dimensions of length, 105 mm; width, 47 mm; height, 3.2 mm. Typical discharge and cycle-life profiles for the Ultralife cells are given in Figure 10.8(a) and (b), respectively. The specific energy of these solid-polymer cells is ~150 Wh kg^{-1}, *i.e.* rather greater than that of liquid-electrolyte cells. The cells may be used from -20 to $+60\,°C$ and have a low self-discharge rate. It will be of considerable interest to see the degree to which solid-polymer lithium-ion cells replace liquid-electrolyte cells for powering portable electronics applications. Indeed, in 1998, Yuasa announced plans to manufacture lithium–polymer cells for notebook-style computers at the rate of 50 000 per month.

10.6   HIGH-TEMPERATURE LITHIUM BATTERIES

In Chapter 7 (Section 7.4), we described a high-temperature *primary* lithium battery, a so-called 'thermal battery', which is used as a power source in military weapons. Here, we consider *rechargeable* lithium batteries which operate at high temperature.

Why the interest in batteries which operate only above 400 °C? What possible applications are there for them? As noted in the introduction to

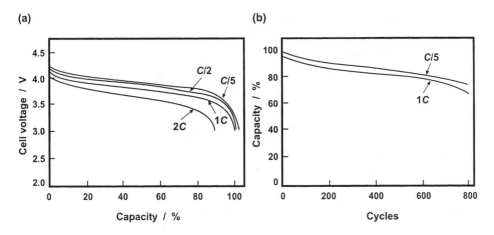

Figure 10.8 *Performance of Ultralife solid-polymer cells at 22°C: (a) capacity profile and (b) cycle-life profile at various discharge rates (By courtesy of Ultralife Batteries, Inc.)*

Chapter 6, one of the potential electrolytes for use with lithium is a fused salt. Inorganic salts generally melt at high temperatures, although there are some, rather complex, salts which melt close to room temperature. To date, these have found no place in practical lithium batteries. The attraction of fused salts as electrolytes lies in their generally high ionic conductivities (100 to 500 S m^{-1}), which leads to the possibility of very powerful batteries indeed, with rapid and reversible electrode reactions. Against these attractions must be set the rather severe materials science and compatibility problems that are created by operating batteries at elevated temperatures.

As regards applications, it is clear that a high-temperature battery will never be suitable for small-scale use because its thermal capacity will be too small and its heat losses too high. Similarly, it is most unlikely to be suitable for consumer use. Rather, high-temperature batteries should be considered candidates for very large power sources where these objections do not arise. Most of the work to date on such batteries has been in the context of energy storage in the electricity-supply system, large UPS units, and traction batteries for road vehicles and submarines. In principle, such batteries may be used in units of, say, 10 kWh upwards.

Early research in the 1960s using fused-salt electrolytes focused on such exotic electrochemical couples as lithium–chlorine and lithium–sulfur at temperatures of 300 to 500°C. Although the theoretical specific energy and anticipated power output of these couples are exciting, it was soon realized that the practical difficulties of juggling two liquids and a gas, or three liquids, in a cell were horrendous, not to mention the problem of

achieving good compatibility between the materials. The research was
soon abandoned in favour of the more modest goal of developing a
lithium–iron-sulfide cell in which the positive electrode, at least, is solid at
the operational temperature. Investigations were carried out on this
battery for many years in the USA, at the Argonne National Laboratory,
with limited programmes also in Germany and the UK.

The fused-salt electrolyte employed in the lithium–iron-sulfide cell was
a binary or ternary eutectic halide. The binary eutectic LiCl : KCl
(melting point: 352 °C) was widely employed in the 1970s and 1980s.
Because the composition of the electrolyte changes during cell use it was
necessary to operate the cells at 450 to 500 °C. In an attempt to lower the
temperature, and so ease corrosion problems, ternary eutectics such as
LiF : LiCl : LiBr and LiF : LiCl : LiI were employed. Argonne preferred
LiCl : LiBr : KBr which could be employed safely at 400 to 425 °C.

There are two sulfides of iron, namely, FeS_2 and FeS. The former
compound discharges electrochemically in two steps, and gives a discharge curve with two, quite distinct, voltage plateaux. Only the upper
one of these is readily reversible. Most work therefore focused on FeS
which is simpler. The reversible discharge reactions are:

$$2Li + 2FeS \xrightleftharpoons[\text{Charge}]{\text{Discharge}} Li_2FeS_2 + Fe \quad V^\circ = +1.64\,V \qquad (10.6)$$

$$2Li + Li_2FeS_2 \xrightleftharpoons[\text{Charge}]{\text{Discharge}} 2Li_2S + Fe \quad V^\circ = +1.62\,V \qquad (10.7)$$

Strictly speaking, FeS also gives two voltage plateaux, but these are so
close together (*viz.* 1.64 and 1.62 V *vs.* lithium at 450 °C) that they may
only be determined by equilibrium thermodynamic measurements. From
a practical point of view, therefore, a Li–FeS cell will show a single
voltage plateau on discharge.

It was soon found that the control and compatibility problems associated with the use of liquid lithium could be greatly ameliorated by using
a solid lithium electrode. This could be done by utilizing a lithium alloy,
which is solid at the operating temperature of the cell and through which
lithium diffuses sufficiently rapidly. The two alloys chosen were LiAl and
Li_4Si. Each alloy has its advantages and disadvantages, although most
research was performed with LiAl. The use of this alloy reduces the cell
voltage by 0.3 V, a penalty well worth paying for not having to contend
with liquid lithium. The result is that the LiAl–FeS cell is a low-voltage
cell with a working value of 1.2 V, *i.e.* comparable with nickel–cadmium

and nickel–metal-hydride.

Many materials-science problems of a compatibility and corrosion nature were encountered in developing the LiAl–FeS cell. A particular difficulty was finding a suitable material for the separator. In the USA, a felt separator made of boron nitride was used, as this is compatible with the electrolyte. In the UK, it was found to be cheaper to employ powdered magnesia as an immobilizer to hold the molten salt electrolyte. Before cell assembly, the magnesia and the halide electrolyte mixture were compressed into plates which were lightly sintered to form plaques. The plaques were then pressure-bonded to the electrodes. The magnesia served as both a separator and an absorber for the electrolyte and was used in the 'electrolyte-starved' mode, *i.e.* in a manner similar to that of the AGM separator in a valve-regulated lead–acid battery (see Section 8.7, Chapter 8).

Both in the USA and the UK, many flat-plate cells were constructed in a dry-room atmosphere and put on test. The cells operated well and gave an excellent, horizontal discharge curve at 1.2 V. The construction of the prismatic LiAl–FeS cell developed in the USA is shown in schematic form in Figure 10.9. The cell has dimensions of length, 130 mm; width, 20 to 40 mm; height, 130 mm. The capacity is ∼90 Ah and the specific energy is 100 Wh kg^{-1}. Disappointingly, the peak power is only 80 W kg^{-1}. Cells constructed by Varta Batterie AG in Germany gave similar performance and it was concluded that unipolar cells would not provide the desired

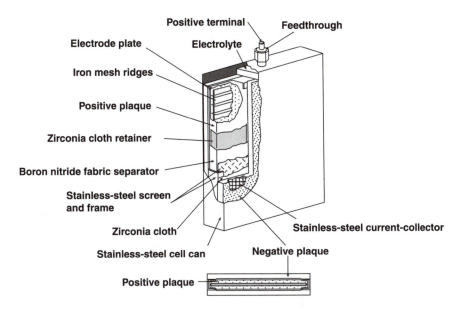

Figure 10.9 *Schematic diagram of prismatic LiAl–FeS cell (USA, 1970s)*

power output. To overcome the problem, bipolar cells were designed and manufactured in the USA and a cell stack was assembled. This gave an improved specific energy of 130 Wh kg^{-1} and a much enhanced specific power of 240 W kg^{-1}. When using FeS_2 for the positive electrode, it was possible to extend the performance to 180 Wh kg^{-1} and 400 W kg^{-1}.

Despite years of research and development, now abandoned in Germany and the UK, if not also in the USA, there are still technical problems with the lithium–iron sulfide battery. The basic principles are now established, however, and it seems likely that a practical and commercially viable battery could be manufactured if the incentive existed.

Chapter 11

Advanced Rechargeable Batteries and Capacitors

Most of the batteries discussed in the foregoing chapters are commercially available; the main exception lies with some of the lithium batteries. In this chapter, we discuss various other types of rechargeable battery which are novel in concept. All are in an advanced state of development, but not all are yet in commercial production.

11.1   ZINC–AIR BATTERIES

The zinc–air battery is novel in two respects: (i) it has a gaseous positive active-material, *viz.* oxygen; (ii) although not electrically rechargeable, it may be recharged 'mechanically' by replacing the discharged product, zinc hydroxide, with fresh zinc electrodes. With these features, the battery is akin to a fuel cell, with the 'fuel' being zinc metal. Actually, it does not fall neatly into the usual definitions of either a fuel cell (two gaseous reactants), a primary battery (single use, throwaway), or a secondary battery (electrically rechargeable). Conventionally, we shall treat it as the latter.

In principle, a zinc–air battery is electrically rechargeable but, in practice, years of research have failed to come up with a viable secondary battery. The problem is the same as that encountered with the nickel–zinc battery (see Section 9.5, Chapter 9), namely, the solubility of zinc hydroxide in potassium hydroxide solution (as zincate ion, $ZnO_2{}^{2-}$). This gives rise to difficulties in achieving uniform electroplating of zinc during recharge, and to the formation of zinc dendrites. In the absence of a technical solution, the idea was conceived of recharging the battery 'mechanically' by removing the spent slurry of zinc oxide and zinc hydroxide suspended in potassium hydroxide electrolyte, and replacing it with fresh electrolyte and fresh zinc electrodes. The slurry is then returned to a central facility for reprocessing *via* the electrical deposition of fresh zinc powder which is recycled to the battery. Thus, the battery may be regarded as an 'indirect secondary battery' as the zinc electrode is re-charged electrically, but not within the battery housing. Although born of necessity, the mechanically-rechargeable battery does have two positive features: (i) it does not require the development of a bi-functional oxygen electrode (*i.e.* one which may be used in charge as well as discharge), which is a difficult task; (ii) recharging is reasonably rapid. Metal–air batteries are also attractive because air is free and does not have to be carried around. These factors improve the prospects for a battery with high specific energy.

Several companies have worked on mechanically-rechargeable batteries. The most advanced technology is that developed in Israel by Electric Fuel Limited. The system has been designed and demonstrated as a traction battery for electric vehicles. The battery has a bed of zinc powder that is compacted onto a current-collector frame and inserted in a separator envelope which is flanked on both sides by air electrodes. The electrode assembly is in the form of a cassette which may be removed from the cell stack for recycling and replaced by a fresh cassette (Figure 11.1). The depleted cassettes are conveyed to a central facility for electrochemical regeneration of particulate zinc. The overall system is composed of three units: the cell stack, the stack dismantling and refuelling machine, and the factory-based plant for fuel regeneration. It has been necessary to develop each unit separately. The overall concept is illustrated in Figure 11.2. Battery modules contain 66 cassettes (unit cells). A traction battery of eight modules (150 kWh) was built and tested success-

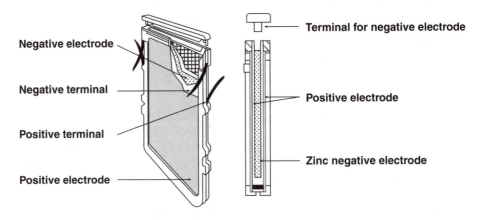

Figure 11.1   *Zinc–air cell and cassette system*
(By courtesy of Batteries International)

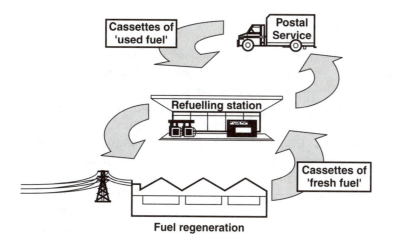

Figure 11.2 *Schematic representation of the operation of electric vehicles (e.g. post office vans) powered by mechanically-recharged, zinc–air batteries*
(By courtesy of Electric Fuel Limited)

fully in Germany in a Mercedes Benz 410 postal van. The battery had high specific energy (200 Wh kg^{-1}), but only modest specific power (100 W kg^{-1} at 80% DoD). The range of the vehicle was 300 km between recharges.

Another version of the zinc–air battery is being developed by Metallic Power in California, USA. In this design, zinc cassettes are refuelled *in situ* by means of a suspension of zinc particles in fresh potassium hydroxide solution.

Other metal–air systems have been investigated and developed to some degree as traction batteries. The iron–air battery was brought to the demonstration stage in Sweden, Germany and the USA, but was found to have serious drawbacks and work has now been discontinued. Similarly, there was a research programme on mechanically-rechargeable, aluminium–air batteries at the Lawrence Livermore Laboratory in the USA, but this too met severe obstacles and was discontinued.

11.2 ZINC–BROMINE BATTERIES

Bromine is another excellent chemical oxidizing agent which could, in principle, be used as a positive active-material. It is a dense, but highly volatile, liquid (boiling point: 58.8°C) which is corrosive, poisonous if inhaled, and causes severe blistering of the skin. Handling requires care. Also, the reactivity of bromine is such that battery design is likely to encounter serious materials' compatibility problems, especially in the area of containment (seals, gaskets, *etc.*). Despite these difficulties, considerable research and development has been carried out on the zinc–bromine battery with a good measure of success.

The zinc–bromine battery was first patented in the 1880s. For almost a century, however, it remained no more than an interesting concept because of its high self-discharge rate, and because of internal short-circuits caused by zinc dendrites. In the 1970s, a number of research groups revisited the system and conducted studies aimed at overcoming these two problems. The most successful approach, of several tried, was to use an acid electrolyte (pH = 3) of zinc bromide ($ZnBr_2$) from which zinc may be electroplated on recharge. The use of a flowing electrolyte resulted in a satisfactory morphology for the zinc deposit. The half-cell reactions are as follows.

At the positive electrode:

$$Br_2 + 2e^- \underset{\text{Charge}}{\overset{\text{Discharge}}{\rightleftarrows}} 2Br^- \quad E^\circ = +1.065 \text{ V} \tag{11.1}$$

At the negative electrode:

$$Zn \underset{\text{Charge}}{\overset{\text{Discharge}}{\rightleftarrows}} Zn^{2+} + 2e^- \quad E^\circ = -0.763 \text{ V} \tag{11.2}$$

Thus, the electrolyte ($ZnBr_2$) is regenerated on discharge. The standard voltage of the cell is 1.83 V. At a current density of 100 mA cm^{-2}, the voltage typically falls to 1.3 V.

The zinc–bromine cell utilizes a bipolar electrode. During charging, bromine is liberated on the positive face of this electrode and zinc is deposited on the negative face. There is a microporous plastic separator through which ions pass to form the discharge product of zinc bromide. Cells are series-connected *via* their bipolar plates. The resulting battery has a positive-electrode loop and a negative-electrode loop (Figure 11.3) and is an example of a 'flow battery'. The soluble zinc bromide formed on discharge is stored, along with the rest of the electrolyte, in the two loops and external reservoirs. The electrolyte in each storage reservoir is pumped through the appropriate loop. With such a battery, as with a fuel cell, the energy-storage capacity of the battery is determined by the size of the storage tanks and their inventory of chemicals, while the power output is determined by the size of the cell stack itself. Thus, the energy and power

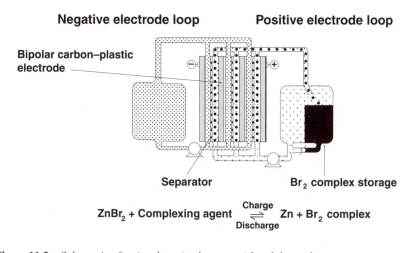

Figure 11.3 *Schematic of a zinc–bromine battery with polybromide storage* (By courtesy of Batteries International)

functions can be separated and sized individually to meet the demands of the intended application. Note, conventional batteries do not have this design flexibility as both the energy-storage capability and the power rating are intimately related to the size and shape of the electrodes.

Several different manifestations of the zinc–bromine battery have been proposed, the most successful of which stores the excess bromine external to the cell as a chemical complex formed with a quaternary ammonium bromide, such as an alkyl morpholinium bromide or an alkyl pyrrolidium bromide. The complex of these salts with bromine is a dense, oily liquid which is immiscible with water. When bromine is liberated during charge, the polybromide complex is produced as droplets. These separate from the aqueous electrolyte stream and are stored in a liquid reservoir external to the cell stack (Figure 11.3). On discharge, bromine is returned to the cell in the form of a dispersion of the polybromide oil in the aqueous electrolyte. This is in equilibrium with dissolved bromine and polybromide ion (Br_3^-), which are both discharged at the positive electrode. The battery does require periodic deep discharge to equalize the state-of-charge of the two electrodes, but suffers no harm from such duty.

In principle, the construction of the cell stack and housing is straightforward. The bipolar plate is a lightweight, carbon–plastic composite which is formed by extrusion. The microporous separator is contained within an injection-moulded, plastic frame which incorporates channels for electrolyte flow. These stack components, which are lightweight and simple to manufacture, are assembled in a 'plate-and-frame' configuration, just like a fuel cell. The plastic has to demonstrate good stability towards bromine. Similarly, it is essential for the reservoirs, connecting tubing, pumps and gaskets to withstand degradation by bromine. The ability to fabricate cell stacks from plastic materials is considered to be a positive factor for the zinc–bromine battery in terms of cost and ease of assembly. Most commercial plastics, however, contain inorganic fillers which are incompatible with bromine over an extended time span, and it may therefore be necessary to use specially prepared plastics. Given these problems with cell materials, it is not surprising that some concern has been expressed about the safety aspects of the battery, particularly for electric-vehicle applications in warm climates.

One of the generic problems with the bipolar cell stacks used in flow batteries is the occurrence of 'shunt currents' through the electrolyte manifolds (note, the manifolds connect electrodes at different potentials). Under the action of such currents, zinc is gradually transferred from electrodes at high potentials to those at low potentials. Various expedients have been adopted to minimize this effect with the result that energy losses associated with shunt currents can now be reduced to $\sim 3\%$ of the discharge energy.

Despite the above difficulties, considerable success has been achieved with zinc–bromine batteries in several countries. For example, a 22 kWh unit was constructed and used to power an electric van operated by the Austrian Post Office. This 108 V battery had a specific energy of 65 to 70 Wh kg^{-1} and a peak specific power of 100 W kg^{-1}. A zinc–bromine battery was also evaluated in an electric racing car. Powercell GmbH of Austria, one of the principal developers of zinc–bromine traction batteries, constructed a 33 kWh, 280 V battery for incorporation in a prototype Daewoo electric vehicle. This battery was claimed to have a specific energy of 80 to 85 Wh kg^{-1} and a life of more than 1000 full discharge cycles. Toyota has also employed such batteries in their electric vehicles. The promising results of these demonstration projects suggest that the zinc–bromine battery should be regarded as a serious contender for a place in future electric transportation, provided residual technical and commercial problems can be solved and safety is not an overriding issue.

11.3 REDOX BATTERIES

Redox batteries are also flow batteries and are constructed on the same principle as the zinc–bromine battery with external reservoirs to store the reactants. In a redox battery, however, there is no electrodeposition of a metal such as zinc and no discrete oxidant such as bromine. Rather, the electrode reactions in the two half-cells involve the oxidation and reduction of ions in the electrolyte; the electrodes are inert and act only as electron-transfer surfaces. An early redox cell was based on the iron–chromium couple, which operates *via* the following electrode reactions.

At the positive electrode:

$$Fe^{3+} + e^- \underset{\text{Charge}}{\overset{\text{Discharge}}{\rightleftharpoons}} Fe^{2+} \quad E° = +0.77\,V \tag{11.3}$$

At the negative electrode:

$$Cr^{2+} \underset{\text{Charge}}{\overset{\text{Discharge}}{\rightleftharpoons}} Cr^{3+} + e^- \quad E° = -0.41\,V \tag{11.4}$$

The cell has a standard voltage of 1.18 V. In the 1980s, iron–chromium redox stacks with power outputs of up to 60 kW were built and tested by Mitsui in Japan as part of a programme to develop energy-storage facilities for electricity utility operations. Further scale-up was abandoned because of the problem of leakage of metal ions through the separator, which lead to cross-contamination of the two half-cells.

A novel way to avoid the problem of cross-contamination is to employ the same metal in different valence states in both half-cells. This is possible with vanadium which exists in four different valence states, namely, vanadium(II) to vanadium(V). The vanadium redox battery has been pioneered by Australian scientists at the University of New South Wales. The half-cell reactions are as follows.

At the positive electrode:

$$VO_2^+ + 2H^+ + e^- \underset{\text{Charge}}{\overset{\text{Discharge}}{\rightleftharpoons}} VO^{2+} + H_2O \quad E° = +1.00\,V \tag{11.5}$$

At the negative electrode:

$$V^{2+} \underset{\text{Charge}}{\overset{\text{Discharge}}{\rightleftharpoons}} V^{3+} + e^- \quad E° = -0.26\,V \tag{11.6}$$

These reactions give a cell with a standard voltage of 1.26 V.

There are four storage tanks external to the battery, *i.e.* two for the reactants in the charged state and two for those in the discharged state. As noted earlier, the energy-storage capacity of such a battery is limited only by the size of the storage tanks. The vanadium redox battery has a low specific energy and a low energy density, but these features may not be too serious for stationary applications in, for example, solar (photovoltaic) electricity systems. Indeed, a 12 kWh battery has been constructed for use in a demonstration solar house in Thailand. The possible advantage of the vanadium redox battery over lead–acid (the traditional choice for solar power systems) lies in its indefinite cycle-life, which is limited only by the materials of construction and not also by the reactants in solution. In Japan, several 200 kWh vanadium redox batteries have been built and are under evaluation for electricity utility operations.

Recently, a new type of redox battery based on the oxidation and reduction of *anions* has been introduced by Innogy plc, a UK electricity utility formerly known as National Power. This is the Regenesys™ battery and utilizes a polymer separator which is permeable to sodium cations but not sulfide anions. During discharge, the reaction at the positive electrode is the reduction of bromine dissolved in sodium bromide (NaBr) solution to bromide ions, while the reaction at the negative electrode involves the oxidation of sulfide ions to sulfur which is contained in sodium polysulfide solution. These reactions are expressed as follows.

At the positive electrode:

$$Br_2 + 2e^- \underset{\text{Charge}}{\overset{\text{Discharge}}{\rightleftarrows}} 2Br^- \quad E^\circ = +1.065\,V \tag{11.7}$$

At the negative electrode:

$$S^{2-} \underset{\text{Charge}}{\overset{\text{Discharge}}{\rightleftarrows}} S + 2e^- \quad E^\circ = -0.508\,V \tag{11.8}$$

Similar to the zinc–bromine and vanadium redox systems, the battery is constructed on the bipolar-plate and frame principle with external storage tanks, one for each electrolyte.

The 'Regenesys™' battery is being developed for large-scale energy storage. Modules in the kW power range have been built and tested (Figure 11.4). Plans are in place to construct a demonstration unit with a storage capacity of 100 MWh at a power rating of 10 MW. The facility will occupy a site area of about one hectare, and thus, will be one of the largest electrochemical storage plants in the world (see Figure 12.4, Chapter 12). Because of its similarity to a fuel cell, in terms of the external

Figure 11.4 *Modules for the 'Regenesys^{TM}' redox flow battery*
(By courtesy of Innogy plc)

storage of chemicals, Innogy plc is promoting the system as a 'regenerative fuel cell' rather than as a battery.

11.4 NICKEL–HYDROGEN BATTERIES

The nickel–hydrogen battery is an alkaline electrolyte system developed in the late-1970s specifically for use in satellites. It resembles the zinc–air battery in that it has one active species which is gaseous, but in this case it is the negative reactant rather than the positive. Both systems are battery–fuel-cell hybrids (see Chapter 9). The practical difference between them is that hydrogen is not readily available and is stored in the battery, whereas air is free and does not have to be transported (at least for terrestrial applications). Also, nickel–hydrogen batteries are electrically rechargeable with a long cycle-life, whereas zinc–air counterparts are not.

There are relatively few negative electrode materials which can be used in strong alkaline solution; the most promising are cadmium, iron, zinc, and hydrogen. Cadmium, iron and zinc batteries with nickel-oxide positive electrodes have been discussed in Chapter 9, along with the nickel–metal hydride battery which grew out of earlier work on the nickel–hydrogen system. The nickel–hydrogen battery has a sintered, nickel-oxide positive electrode and a platinum–hydrogen negative electrode. The battery was developed specifically to replace nickel–cadmium in space applications on account of its somewhat higher specific energy (~ 50 Wh kg^{-1}) coupled with a very long life when subjected to deep-discharge cycling.

The hydrogen electrode is fully reversible and, when combined with a nickel-oxide positive, provides a cell with a standard voltage of 1.32 V. The half-cell reactions are as follows.

At the positive electrode:

$$\text{NiOOH} + \text{H}_2\text{O} + \text{e}^- \underset{\text{Charge}}{\overset{\text{Discharge}}{\rightleftarrows}} \text{Ni(OH)}_2 + \text{OH}^- \quad E^\circ = +0.490 \text{ V} \quad (11.9)$$

At the negative electrode:

$$\tfrac{1}{2}\text{H}_2 + \text{OH}^- \underset{\text{Charge}}{\overset{\text{Discharge}}{\rightleftarrows}} \text{H}_2\text{O} + \text{e}^- \quad E^\circ = -0.828 \text{ V} \quad (11.10)$$

Note that there is a counterflow of water molecules and hydroxide ions across the separator. The hydrogen gas liberated on charging is stored under pressure within the cell itself. This has necessitated considerable development work on the design and construction of the required pressure vessel, on the electrical leads through the vessel wall to the battery, and on the cell stack and its supports. The pressure vessel is cylindrical in shape, with hemi-spherical end caps, and is constructed of thin-gauge, Inconel alloy (Figure 11.5). Cycling of the cell is accompanied by a change in the hydrogen pressure, *i.e.* from ~4 MPa in the charged state to ~0.2 MPa in the discharged state. The pressure vessel must withstand this treatment without embrittling, corroding or cracking.

The cell stack is built up from back-to-back sintered nickel electrodes, which are electrochemically impregnated, and standard hydrogen fuel-cell electrodes which consist of platinum black dispersed on carbon paper. The separators are formed from a porous ceramic paper, made from fibres of yttria-stabilized zirconia, which absorbs the potassium hydroxide electrolyte. The cells may be overcharged since oxygen liberated at the positive electrode recombines rapidly at the negative electrode. (Note, at very high rates of overcharge, heat must be transferred away from the cell to avoid thermal runaway.) Nickel–hydrogen space-cells are sophisticated products which are exceedingly expensive to buy, but which nevertheless have largely usurped nickel–cadmium for use in satellites (see Section 12.2, Chapter 12).

11.5 SODIUM–SULFUR BATTERIES

Sodium has many attractions as a negative-electrode material. It has a high reduction potential of -2.71 V (compare, for example, zinc at

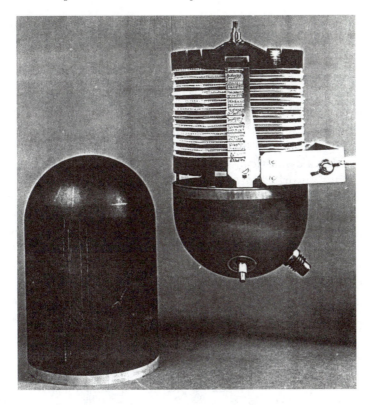

Figure 11.5 *Nickel–hydrogen cell (40 Ah) as used in satellites*
(By courtesy of AEA Technology Batteries)

−0.76 V) and a low atomic mass (23.0). Together, these properties offer
the prospect of a battery with high specific energy. Moreover, sodium
salts are abundant in nature and non-toxic, and the metal is readily
extracted and comparatively cheap. Of the possible positive-electrode
materials which can be used in combination with sodium to form a cell,
sulfur is superficially the most attractive. It, too, is readily available and
very cheap, almost to the point of being a waste product.

The problem of devising a conceptual sodium–sulfur cell reduces to
that of finding a suitable electrolyte. Clearly, aqueous electrolytes cannot
be used and, unlike the case of lithium (see Section 10.5, Chapter 10), no
suitable polymer is known. Fortunately, there is an alternative. In 1966,
scientists working at the Ford Motor Company in the USA discovered a
ceramic material – beta-alumina – which is an electronic insulator, but
which has a high ionic conductivity for sodium ions above 300 °C. Beta-
alumina is a sodium aluminium oxide with a rather complex crystal
structure. Indeed, there are several forms of beta-alumina with somewhat

different compositions and structures. The highest ionic conductivity is given by β''-Al_2O_3 which has an idealized composition of $Na_2O \cdot 5.33Al_2O_3$. At 350°C, the conductivity for sodium ions is as high as $20\,S\,m^{-1}$, a value comparable with that of many aqueous electrolytes.

The Ford scientists went on to develop and demonstrate sodium–sulfur cells based on beta-alumina electrolyte in the form of ceramic tubes. In each cell, the molten sodium (negative electrode) was contained in a vertical closed-end tube, of diameter 1 to 2 cm, which was housed within a cylindrical steel case. The positive electrode consisted of molten sulfur absorbed into the pores of carbon felt (the current-collector) and packed into the annulus between the ceramic tube and the steel case. Subsequently, this concept was taken up by numerous developers in Europe and in Japan. A schematic of a 40 Ah cell produced by Asea Brown Boveri GmbH in Germany is shown in Figure 11.6. In this version, the steel container has been replaced by one made from aluminium for lightness.

The cell discharges at 300 to 400°C in two steps as sodium ions (formed by ionization of the metallic sodium) pass from the negative electrode, through the beta-alumina, to the sulfur positive electrode, *i.e.*

step 1:

$$2Na + 5S \underset{Charge}{\overset{Discharge}{\longleftarrow}} Na_2S_5 \quad V° = +2.076\,V \qquad (11.11)$$

step 2:

$$2xNa + (5-x)Na_2S_5 \underset{Charge}{\overset{Discharge}{\longleftarrow}} 5Na_2S_{5-x} \quad V° = +2.076 \rightarrow +1.78\,V \quad (11.12)$$

In the first step, sodium polysulfide (Na_2S_5) is in equilibrium with sulfur, in the form of two immiscible liquids, and thus the voltage is constant. Once all the sulfur has been converted into Na_2S_5, the second step of the reaction involves the production of lower polysulfides. This is a homogeneous, single-phase reaction and the voltage declines linearly to 1.78 V at the composition Na_2S_3. At this point, the discharge is stopped to avoid the formation of insoluble Na_2S_2.

Although simple in concept, the sodium–sulfur battery proved to be difficult to develop in practice. Large teams of scientists and engineers were set up in Germany, Japan, the UK and the USA to bring the battery to commercial realization. The research continued for up to 25 years. The major problem areas proved to be the following.

Beta-alumina electrolyte. Beta-alumina is a highly sophisticated techni-

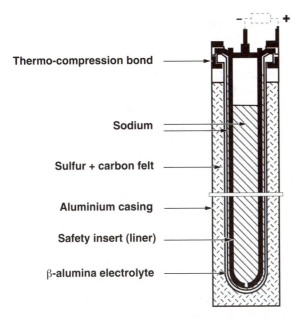

Figure 11.6 *Schematic cross-section of Na–S cell (40 Ah) developed by Asea Brown Boveri GmbH*
(By courtesy of Asea Brown Boveri GmbH)

cal ceramic and is required to meet exacting specifications. Aside from the electrical properties, which are determined by the composition of the ceramic, there is a requirement for near theoretical density, small grain size, and good mechanical properties (tensile strength, fracture toughness, *etc.*). Additionally, there are stringent demands on fabrication technology to mass-produce tubes of uniform quality, straightness and dimensional tolerance, and with reproducible properties. All this has been achieved by several manufacturers in the above-mentioned countries.

Sealing of cells. Cells have to be sealed to separate the sodium and sulfur compartments, not only electrically but also physically to prevent transport of vapour from one compartment to the other and to prevent ingress of air. The seals have to be resistant to chemical attack by both sodium vapour and sulfur vapour, and be capable of thermal cycling without cracking. These requirements present a very difficult challenge. After much trial and error, most manufacturers eventually settled for an insulating glass collar at the top of the cell to which suitable metallic components were joined by a thermo-compression bond.

Corrosion of cell cases. Sodium polysulfide is highly corrosive towards metals, especially steel. To overcome this problem, special liners and coatings of high-chrome alloys had to be developed.

Sodium purity. To prevent embrittlement and subsequent fracture of the ceramic tube, it is important to use sodium of high purity. In laboratory work, it was usual to purify the sodium and fill the cells in argon-filled glove-boxes. Later, engineered vacuum lines were constructed for dispensing liquid sodium.

Safety. The chemical reaction of molten sodium and sulfur is violent and leads to an uncontrolled fire. Early experience showed that the fracture of the electrolyte tube caused a fire inside the cell and, consequently, corrosion and puncture of the cell housing. This was overcome by incorporating safety devices in the cell. Most designs have taken the form of a liner to the beta-alumina tube (to wick up the sodium) with a pin-hole in its base, as shown in Figure 11.6. This is adequate to allow a normal flow of sodium to the inner wall of the electrolyte tube, but restricts the flow in the event of tube fracture.

In addition to these cell problems, two difficulties were encountered in assembling cells into batteries. First, for the batteries to be thermally self-sustaining, at temperatures between 300 and 400 °C, it was necessary to develop an insulated housing and a thermal control system. Such a requirement dictated that the battery had to be of a minimum size of several kWh. Consequently, the battery has been developed solely for electric vehicle and stationary storage applications. The second difficulty relates to the absence of an overcharge mechanism which can cause one or more cells to develop high resistance when fully charged so that there is no mechanism for balancing units in a series chain. When this occurs in a sodium–sulfur cell, the entire voltage of the series chain effectively falls across the cell at top of charge and results in dielectric breakdown of the ceramic tube. Moreover, a failed cell generally goes to open-circuit and, thereby, isolates all others in series with it. The problem was solved by means of a cell by-pass arrangement, which activated when a cell failed, and by configuring the battery as a series–parallel array of cells with parallel connections after every few cells in a series chain. Such a system limits the number of cells lost when one cell fails to open-circuit.

Despite the above technical difficulties, remarkable success was achieved with sodium–sulfur batteries. Vehicle traction batteries were built and tested, especially in Germany and the UK. For example, a 10 kWh battery module, manufactured by Asea Brown Boveri GmbH, is shown in Figure 11.7 with part of the housing cut away to reveal the cells inside. The module is fitted with leads for mains power supply, voltage monitoring, heating and earth protection. The specific energy is ~ 100 Wh kg^{-1}. Many hundreds of the modules were produced up until 1993, and traction batteries were installed in electric cars and vans. By November 1992, the cars had covered a total of 600 000 km without any

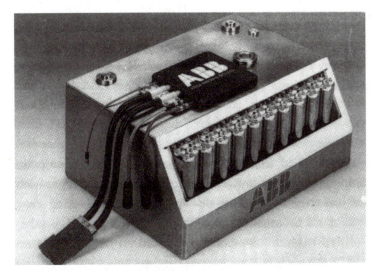

Figure 11.7 *Na–S battery module (10 kWh) developed by Asea Brown Boveri GmbH* (By courtesy of Asea Brown Boveri GmbH)

kind of accident attributable to the batteries. Nevertheless, the development programme was eventually terminated due to a combination of technical and commercial reasons.

A somewhat different approach was adopted in the UK. After trying many different designs, the developer (Chloride Silent Power Limited) eventually settled on a small cell of diameter and height both 45 mm, and capacity 10 Ah. These cells are shown schematically in Figure 11.8, and photographically in Figure 11.9. Small cells were chosen for technical and safety reasons, but it is hard to see how they would be economic to manufacture. The cells had a specific energy which approached 200 Wh kg^{-1} at the 3-h rate. A pilot plant to manufacture 250 000 cells per year was built in the early 1990s, but eventually this initiative ceased also. Sodium–sulfur batteries are no longer being considered for traction applications (so far as the authors are aware), although the Tokyo Electric Power Corporation in Japan are advancing megawatt-sized units for energy storage in electricity supply networks.

What of the future? Although much success was achieved and many of the problems were solved, it seems unlikely that the sodium–sulfur battery will be taken up again in Europe and/or the USA in the near term. The fundamental obstacle lies in the absence of an overcharge mechanism for the cell. Although it is possible, to some extent, to engineer a way around this limitation, the procedure will never be fully satisfactory. In

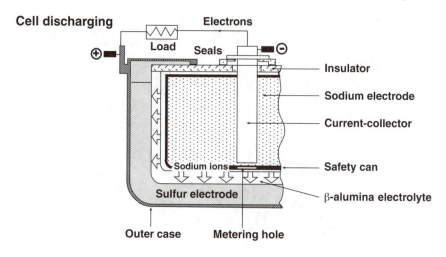

Cell discharging

Figure 11.8 *Schematic of Na–S cell (10 Ah) developed by Silent Power Limited. Cell in discharge mode*
(By courtesy of Silent Power Limited)

Figure 11.9 *Na–S cells (10 Ah) developed by Silent Power Limited*
(By courtesy of Silent Power Limited)

fact, interest in high-temperature sodium batteries has moved towards the sodium–metal-chloride system which is simpler in concept and easier to manufacture.

11.6 SODIUM–METAL-CHLORIDE (ZEBRA) BATTERIES

The sodium–metal-chloride battery is a derivative of the sodium–sulfur battery in which the sulfur positive electrode is replaced by nickel chloride or by a mixture of nickel chloride ($NiCl_2$) and ferrous chloride ($FeCl_2$).

The sodium negative electrode and the beta-alumina ceramic electrolyte are the same as in the sodium–sulfur battery. The basic cell reactions are simple, *i.e.*

$$NiCl_2 + 2Na \underset{\text{Charge}}{\overset{\text{Discharge}}{\rightleftarrows}} Ni + 2NaCl \quad V° = +2.58\,V \tag{11.13}$$

$$FeCl_2 + 2Na \underset{\text{Charge}}{\overset{\text{Discharge}}{\rightleftarrows}} Fe + 2NaCl \quad V° = +2.35\,V \tag{11.14}$$

The sodium–metal-chloride battery was conceived in the early 1980s through collaboration between scientists working in the UK and in South Africa. It became known as the ZEBRA battery. Many of the research team had previously worked on sodium–sulfur batteries, so progress was comparatively rapid. By 1984, the first ZEBRA-powered electric vehicle was being driven around the streets of Derby in the UK.

Early attempts to assemble cells in the charged state were not successful. It was soon found, however, that the way forward was to assemble cells in the discharged state with the positive electrode pre-formed from a mixture of metal powder (nickel or iron) and common salt (sodium chloride, NaCl). On charging the cell, these chemicals were converted into the corresponding metal chloride in the positive-electrode compartment and sodium in the negative-electrode compartment. This procedure gave the following benefits: (i) it was no longer necessary to handle liquid sodium, which is formed *in situ*; (ii) the sodium prepared by diffusion through beta-alumina is ultra-pure; (iii) the raw materials for the battery (metal powder and common salt) are inexpensive. It was further found to be advantageous to construct the cell with a central positive electrode, *i.e.* with the solid starting materials inside the electrolyte tube and the sodium generated in the annulus between the beta-alumina tube and the cell case. The latter could then be made of mild steel, a cheap commodity, which is compatible with molten sodium.

Experiments were conducted with both ferrous chloride and nickel chloride electrodes. Although iron powder is cheaper than nickel powder, it was discovered that nickel cells presented fewer complications and could be operated over a wider temperature range (200 to 400°C) than iron cells (200 to 300°C). Consequently, attention was focused on nickel-based cells.

A problem arises at the positive-electrode | electrolyte interface, namely, both components are solids and the contact between them is inadequate to sustain a high current flow. This restriction on performance was

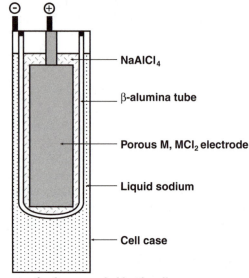

Figure 11.10   *Schematic of sodium–metal-chloride cell*
(By courtesy of Research Studies Press Ltd)

overcome by introducing a second electrolyte, a liquid, to make good
ionic contact between the two solids. Molten sodium chloraluminate
($NaAlCl_4$) was chosen as the electrolyte. When the cell is assembled, the
molten salt is simply poured into the positive-electrode compartment and
impregnates the porous compact of nickel and sodium chloride. Excess
liquid forms a narrow annulus between the electrode and the inner wall of
the beta-alumina tube. The molten $NaAlCl_4$ is a vital component, but its
extra mass does lower the specific energy of the ZEBRA battery (by
~ 10%). A schematic of the sodium–metal-chloride cell is given in Figure
11.10, and a selection of pre-formed positive electrodes of varying capacity is shown in Figure 11.11.

   One of the advantages of the sodium–nickel-chloride cell over the
sodium–sulfur equivalent is that it has intrinsic provision for both an
overcharge and overdischarge reaction. At the top-of-charge, when all the
solid sodium chloride has been decomposed, overcharge results in depletion of sodium ions in the melt and further chlorination of the porous
nickel matrix, *i.e.*

$$2NaAlCl_4 + Ni \quad \underset{\longleftarrow}{\overset{\text{Overcharge}}{\longrightarrow}} \quad 2Na + 2AlCl_3 + NiCl_2 \qquad (11.15)$$

Overdischarge also involves a reversible reaction with the sodium

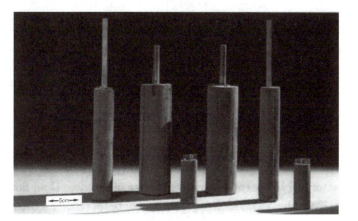

Figure 11.11 *Pre-formed positive electrodes of nickel powder and sodium chloride (10, 60 and 100 Ah capacity)*
(By courtesy of AEA Technology Batteries)

chloraluminate, namely:

$$3Na + NaAlCl_4 \underset{\xleftarrow{\hspace{1cm}}}{\overset{\text{Overdischarge}}{\xrightarrow{\hspace{1cm}}}} Al + 4NaCl \tag{11.16}$$

These reactions are shown diagrammatically in Figure 11.12. The importance of these overcharge and overdischarge reactions is that it is possible to couple cells in a single series chain, without the need for parallel connections, and to balance cells of slightly different capacity.

The second major advantage of the sodium–nickel-chloride system lies in the area of safety. Much experience from extensive testing has shown these cells and batteries to be safe under almost all conditions of use and abuse. For example, if the beta-alumina electrolyte tube cracks, the molten sodium first encounters the $NaAlCl_4$ electrolyte and reacts with it as shown in Equation (11.16). The resulting aluminium powder ensures that the cell fails to short-circuit rather than to open-circuit, while the solid sodium chloride tends to plug the crack. In a series chain of cells, the result is simply the loss of one 2 V unit and the battery continues to operate.

Several thousand cells of 40 Ah capacity were made and assembled into batteries. Details of the cell design changed as the work progressed. For instance, later versions employed cell cases with a square cross-section so as to improve the packing density in the thermal enclosure and thus maximize the capacity of the battery. Overhead views of these closely-packed cells are shown in Figure 11.13; the cells are electrically isolated

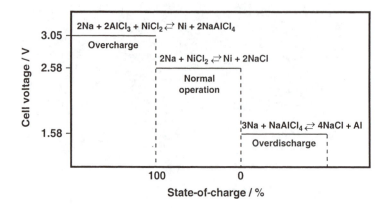

Figure 11.12   *Charge–discharge voltage of sodium–nickel-chloride cell at 250°C*
(By courtesy of Beta Research & Development Ltd)

Figure 11.13   *Two views of ZEBRA cells in mild-steel cans of square cross-section*
(By courtesy of Beta Research & Development Ltd)

from each other by a thin sheet of mica.

A special type of double-walled, evacuated, thermal enclosure made of stainless-steel was constructed as the battery container. The space between the walls contained a lightweight, rigid insulant to prevent buckling or collapse of the thin steel sheets. At the operating temperature of the battery (300 to 400°C), the outer surface skin of the container is only 10 to 20°C above ambient and the heat loss from a vehicle-size pack is between 100 and 300 W.

Batteries of varying size have been constructed to suit different vehicles. A typical battery pack rated at 17 kWh contained 220 cells. The latest batteries are larger, *viz.* 29 kWh. These vehicle batteries have a specific energy of ~ 100 Wh kg^{-1} and an energy density of ~ 160 Wh dm^{-3}; values which are sufficient to meet the required daily range of a city car or van. On the debit side, the battery has a comparatively low specific power. The power is 150 W kg^{-1} at full charge, but declines during discharge. This limitation in performance arises from the thickness of the positive electrode and the associated, marked increase in internal resistance of the cell during discharge. Recent work using shaped (fluted) electrolyte tubes has been directed towards improving the specific power to meet the targets set in the USA for electric-vehicle battery packs (see Table 12.4, Chapter 12), and steady improvements have been reported. Batteries have also been demonstrated to meet the accompanying, mid-term, life targets of at least five years service in the vehicle and 600 charge–discharge cycles at 80% DoD.

ZEBRA batteries have been fitted to a range of electric vehicles, especially the Mercedes-Benz 190E and the Mercedes 'A' cars. These vehicles have a range which exceeds 150 km per charge when driven at up to 70 km h^{-1} (*i.e.* at urban speeds), and individual cars have been in service for many months over substantial distances without any battery accident or breakdown. Indeed, one Mercedes 190E car completed 77 000 km over 2.5 years with no cell failures. The reaction of drivers has been highly favourable. Following a change in ownership of the development company (Beta Research & Development Ltd), a pilot production line for cells and batteries is now being assembled in Switzerland. There is confidence that the ZEBRA battery holds promise as a future power source for electric vehicles. The many attractive features, as seen from the viewpoint of both the manufacturer and the user, are listed in Table 11.1.

The chief disadvantage of high-temperature sodium batteries is that their use is restricted to large units. There is no possibility of entering the market with small consumer units and working up steadily. To tackle the traction battery market head-on is a major undertaking and, in order to meet cost targets, it will be necessary to construct a large and expensive production plant in anticipation of a mass market which does not yet exist.

11.7 SUPERCAPACITORS AND ULTRACAPACITORS

A conventional 'electrostatic' capacitor consists of two conductors ('plates') separated by an insulator (a 'dielectric'). The device stores energy by the separation of positive and negative electrostatic charge; the two plates carry equal but opposite charges. This process is known as non-

Table 11.1 *Manufacturer and user perception of the advantages of ZEBRA batteries*

Manufacturer	User
• Low-cost materials – apart from the nickel, which can be recycled • Assembly in the discharged state – no sodium handling • Flexible cell design • Large cells (up to 500 Ah) are possible • Cells may be close-packed in the battery • No corrosion protection necessary • Failed cells conduct current; this allows series strings with no cross-connections • No added safety features required	• High cell voltage • High specific energy (4 to 5 times that of lead–acid) • 100% coulombic efficiency • Long life (> 1000 cycles); very low rate of cell failure • Wide operating temperature range (nickel-based cells); performance independent of ambient temperature • Sealed and maintenance-free • Overcharge and overdischarge mechanisms • Safety in use fully demonstrated; no volatile components

Faradaic storage of electrical energy. The capacitance, which is the ratio of the magnitude of the charge on either plate to the magnitude of the potential difference between them, is inversely proportional to the inter-plate spacing, and directly proportional to the dielectric constant of the insulating material (vacuum, air, plastic, *etc.*). The stored energy density is very low, typically ~ 0.05 Wh dm^{-3}. The term 'electrolytic capacitor' refers to the well-known, moderately high-capacitance device which is based on a thin-film oxide dielectric formed electrolytically on such metals as aluminium, tantalum, and titanium.

Supercapacitors differ from conventional electrostatic and electrolytic capacitors in that they contain an electrolyte which enables the electrostatic charge to be also stored in the form of ions. Electrodes of high internal surface area – often prepared as compacts of finely-divided, porous carbon – are used to adsorb the ions and to provide a much greater charge density than is possible with non-porous, planar electrodes. The voltage is lower than for a conventional capacitor, while the time constant for charge and discharge is longer because ions move and reorientate more slowly than electrons. In these respects, the supercapacitor begins to take on some of the characteristics of a battery, although no Faradaic (electrochemical) reactions are involved in the charge and the discharge process.

The ultracapacitor moves one step closer to a battery. It stores energy by ionic capacitance as well as by surface (and near-surface) redox processes which occur during charge and discharge. The latter are Faradaic reactions which enhance the amount of stored energy, but which, because they are confined to surface layers, are fully reversible for a long cycle-life. The electrolyte may be either an aqueous solution, *e.g.* sulfuric acid or potassium hydroxide, or an organic solution, *e.g.* tetraethylammonium tetrafluoroborate, $(C_2H_5)_4NBF_4$, in acetonitrile. Capacitors with aqueous electrolytes have a very low resistance, but also a low break-down voltage, while the converse is true for capacitors with organic electrolytes. It should be noted that there is considerable confusion in the literature over the definition of 'supercapacitor' and 'ultracapacitor', especially as groups working on the development of capacitor devices tend to use the terms interchangeably. It has been proposed that the generic name 'electrochemical capacitor' should be adopted.

The performance of batteries and capacitors as energy-storage devices is summarized in Table 11.2. It will be seen that capacitors contrast with batteries in that they store very little energy, but can produce high power output for very short periods. They also have exceptionally long cycle-lives. The contrast is so great that the applications for capacitors and batteries are quite different. Supercapacitors and ultracapacitors are ideal for meeting sudden transient power demands which are too great to be supplied by a battery. The devices are being developed and evaluated for many possible applications. These include electric and hybrid electric vehicles in which peak-power demands are often of short duration. For example, a pack of 104 supercapacitors, which can deliver 50 kW of power for 10 s, has been installed in both the 'aXcessaustralia LEV' and the Holden 'ECOmmodore' hybrid electric cars (Figure 11.14). The pack is combined with a purpose-built lead–acid battery (the CSIRO 'Double-Impact^{TM}' design) to provide a practicable and affordable surge-power unit for vehicle acceleration and hill-climbing. The supercapacitors also provide an extremely efficient means of capturing the high power produced by regenerative braking of the vehicle.

Clearly, electrochemical capacitors are entirely complementary to

Table 11.2 *Comparison of characteristics of energy-storage devices*

Characteristic	Battery	Capacitor		
		Conventional	*Super*	*Ultra*
Energy density (Wh dm^{-3})	50–250	0.05	1–5	1–10
Power density (W dm^{-3})	150	$>10^8$	$>10^3$	$>10^3$
Discharge time (s)	$>4 \times 10^3$	<1	1–10	1–10
Cycle-life	10–1000	$>10^6$	$>10^5$	$>10^5$

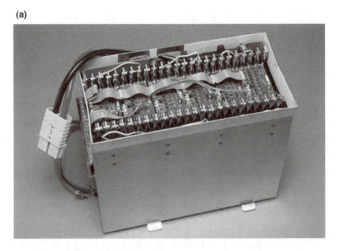

Figure 11.14   (a) *Supercapacitor pack fitted to* (b) '*aXcessaustralia LEV*' *and* (c) *Holden*
'*ECOmmodore*' *hybrid electric cars*
(By courtesy of CSIRO Energy Technology)

batteries. Whether or not they might be of use, alongside batteries, in applications such as portable power supplies and the storage of photovoltaic electricity has yet to be established, but obviously they would be employed only in situations which require transient, high-power pulses.

Chapter 12

Some Applications for
Secondary Batteries

So far, we have given an outline of the various types of battery and some of their principal characteristics. In this final chapter we examine in greater depth the battery specifications set by users for particular applications. Four examples are considered where secondary batteries are integrated into an engineered system. The object is to show how highly interactive the battery and its application can be, and how non-technical factors may also play a significant role in battery selection.

12.1 STORAGE OF SOLAR-GENERATED ELECTRICITY

Applications for photovoltaic ('solar') cells vary greatly both in nature and in size. The size ranges from small single cells of mW power output, as used in solar-powered calculators, through solar modules (panels) designed to supply electricity to caravans, small boats or navigation lights (usually, 36 to 50 cells with a peak power output 50 to 70 Wp), to the much larger arrays of modules used on buildings or for remote-area power supplies (RAPS). These multi-panel arrays have peak power outputs from 1 kWp to over 1 MWp. (Note, the power output of a photovoltaic cell or module is rated in peak-watts, Wp.)

Given that solar energy gain (insolation) is two-dimensional, a system with large power output requires a large area covered by solar cells. For example, the south of England receives an insolation of around 1 TWh per square km per year. With photovoltaic arrays of 15% efficiency, it would be necessary to cover over 2000 square km with cells to generate the UK's electricity consumption of ~ 325 TWh per year! The most elegant and cost-effective method of deploying such area-intensive technology is on the roofs of buildings, rather than as free-standing arrays.

188

Such solar generators could either feed into the main electricity grid or be used to supply electricity locally to the surrounding buildings.

If solar-generated electricity is to be adopted on a large-scale, it is likely first to be through the installation of dispersed smaller units of, say, 1 to 10 kW, although some grid-connected demonstration plants of MW-size have already been constructed. Such plants are by no means economically competitive at present, but the cost of solar modules is predicted to fall substantially in the years ahead. The greatest opportunity for solar-generated electricity is in tropical and sub-tropical regions where insolation is high, evenings are dark, and many small communities are remote from the mains electricity supplies. For this opportunity to be realized in full it will be necessary to have a practical and cost-effective means of storing photovoltaic electricity for hours, and sometimes for days. At present, the best approach is to use secondary batteries, although other options do exist, *e.g.* pumped hydro, flywheels, hydrogen storage.

Secondary batteries have a number of features which make them well suited to storing solar-generated electricity, namely:

- the system input and output is in the form of low-voltage, d.c. electricity;
- batteries respond immediately to supply and load variations, and are very reliable;
- it is possible to match the internal resistance of the battery to that of the load for maximum power output;
- modular construction allows flexible sizing and easy battery exchange;
- batteries have a short lead-time in manufacture.

A schematic of an array–battery combination connected to d.c. and a.c. loads is shown in Figure 12.1.

A solar module consists of a combination of photovoltaic cells in series (to match the battery voltage, often 12 or 24 V). The modules are then connected in parallel to multiply the current and reduce the resistance of the array. The following features determine the choice of battery:

- the output of the array and its variation with time of day and with month of the year as insolation levels change (which gives rise to changes in the battery charging current);
- the environmental conditions, especially temperature;
- the requirement for reliability and freedom from maintenance, as dictated by remoteness and difficulty of access;

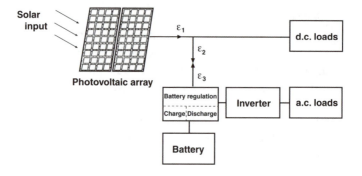

Figure 12.1 *Photovoltaic system to supply solar-generated electricity to d.c. and a.c. loads.* (ε_1, ε_2 and ε_3 = array, charge and discharge efficiency, respectively)

- the pattern of load demand placed on the battery over each charge–discharge cycle;
- the energy efficiency of the battery (*i.e.* Wh output : Wh input);
- the rate of self-discharge of the battery on standing;
- the operational life of the battery;
- the commercially acceptable cost.

Generally, it will be necessary to prioritize these features for any particular application and accept a compromise solution.

The present applications for photovoltaic arrays are quite diverse, but often have in common a remoteness from the mains electricity supply. Examples include the following:

Maritime: Navigation buoys, drilling platforms, cathodic protection of structures, independent electricity supplies for small boats.

Terrestrial: Microwave relay stations, telecommunications, meteorological stations, military installations, railway signalling, street lighting, irrigation, pipeline monitoring and cathodic protection, electric fences, power supplies for remote communities and homesteads, holiday caravans.

Space: Power sources for satellites (a minor use in terms of market volume, but vital for modern telecommunications).

These are all types of RAPS system. As solar cells and modules decline in price, their use in such systems will assume increasing importance. The size of the batteries required to store the solar electricity varies according to the application. Most solar installations are static and for these the mass and volume of the battery will not be prime considerations. Rather, the key parameters are high energy efficiency, long life, good charge retention, reliable maintenance-free operation, and low cost.

Most RAPS applications will lie in the power range 0.1 to 10 kW and will require a battery with an energy-storage capacity of 1 to 100 kWh. By way of comparison, the capacity at the lower end of the range corresponds to that of a large automotive battery, and at the upper end to that of a traction battery pack used in an electric bus. In many locations it will be essential to use a diesel generator and, optionally, a wind generator in conjunction with a solar array, either because the insolation is inadequate or because an array of the required size is too expensive. Such facilities are known as 'hybrid RAPS systems'.

The role of a battery in a RAPS installation depends very much on the insolation level. In a tropical zone, there is little seasonal variation in insolation, and in a desert region there is little rainfall and little cloud cover. Under these conditions, the insolation is fairly predictable and the function of the battery is mainly to store excess electricity generated during the hours of daylight for use during the hours of darkness ('supply levelling'). Only occasionally will it be necessary to store electricity for more than a day and so the extra margin of battery capacity required is quite small. Contrast this with the situation in a temperate zone which has a high occurrence of cloud cover. Because the insolation is now low and erratic, a much larger array is required, as well as a considerable excess of battery capacity to allow for several consecutive days of overcast conditions. An added complication is the major seasonal variation at these higher latitudes. Clearly, for a given application, RAPS systems will be larger, and hence more expensive, in temperate than in tropical zones.

Simple algorithms have been prepared that allow calculation of the daily load (in terms of kW and kWh) for a particular application, and also the weekly averaged load. The latter is particularly important when there is an appreciable daily variation in the use factor. An example of this would be a school which is used only on weekdays, or a holiday caravan occupied only at weekends. The maximum anticipated daily load determines the size of the battery and, by making a simple calculation of this load, it is possible to avoid installing a battery which is too large and expensive. The size of the array is determined by the weekly averaged load, with the battery acting as a 'buffer' between days of low load and those of high load. The commercial objective must be to minimize the overall cost of the complete installation. Much may be done by way of load management, *e.g.* by using low-energy lamps for lighting and well insulated, high-efficiency refrigerators. The extra capital cost of these items is likely to be small compared with the cost of a larger array–battery system.

Seasonal variations in insolation present a greater problem because it will not generally be economic to install extra batteries to provide seasonal energy-storage (here, hydrogen storage may be more practicable). Fortunately, some applications may require electricity only in summer when insolation is at a maximum; examples would be caravans, pleasure boats, or remote holiday cottages. In other situations, the winter load (lighting, television) may differ from the summer load (refrigeration, air-conditioning). Where a RAPS system is required all year round it may be possible to combine solar generation in summer with wind generation in winter since, statistically, winds are stronger and more consistent in winter when insolation is at its lowest.

Another seasonal aspect of RAPS design is to match the tilt angle (to the horizontal) of the array to the load demand. In countries at intermediate latitudes, the maximum insolation in summer falls on an array with a low tilt angle, whereas in winter, when the sun is lower in the sky, maximum insolation is obtained at a higher angle, *i.e.* perpendicular to the sun's rays at solar noon. When using a fixed-angle array (*i.e.* a non-steerable array which does not track the sun), the tilt angle should be chosen so that the insolation profile throughout the year best matches the load requirements. For example, if the load demand is small in winter and large in summer (for air-conditioning or water pumping), then it is best to have a nearly flat array. Alternatively, for loads which are almost constant throughout the year (*e.g.* repeater stations or navigational aids) it is more appropriate to use an inclined array so as to maximize the electricity generated during winter months while sacrificing the excess which is potentially available in summer. By choosing a tilt angle which gives the optimum match, the size and the cost of the array are minimized.

Most batteries, although not all, provide maximum performance when operated at ambient temperature, say 0 to 40 °C. Outside of this temperature range, the performance deteriorates rapidly. This may pose problems in certain locations. Consider, for example, continental countries at northern latitudes (*e.g.* Canada, Russia) where there may be good winter insolation on days when the temperature is -20 to -40 °C. In such places it will be necessary to select a battery which operates well at low temperatures, for instance nickel–cadmium or lithium-ion. A more common problem will be in tropical situations in summer when the temperature of an outdoor battery could easily exceed 70 °C. Batteries which employ aqueous electrolytes will not perform well under these conditions and, therefore, will have to be enclosed in thermally insulated (or ventilated) containers. A possible alternative would be to select a moderate-temperature battery, *e.g.* lithium–polymer, should such technology become commercially available at an affordable price.

A second role for the battery in a solar installation, quite distinct from its day/night storage function, is to accommodate surges in power demand ('peak shaving'). Power surges of relatively short duration and up to six times the steady load are quite routine, as appliances are turned on and off. Without the provision of a battery it would be necessary to have a much larger array which is sufficient to meet the maximum instantaneous demand. This would be uneconomic. Similarly, when using an inverter to provide alternating current for domestic uses, inductive loads (*e.g.* motors in washing machines and vacuum cleaners, compressors in refrigerators) all give rise to short-duration surges of current. Both the battery and the inverter must be capable of meeting these instantaneous loads. While direct-current versions of some of these appliances are commercially available, as used in caravans and boats, they do tend to be more expensive.

The third and final role of the battery is to smooth the swings in current and voltage output from the array. Without this battery function, the power supply to the load would be erratic. Thus, in addition to the primary role of providing a diurnal storage capability, the battery serves as a buffer to match a fluctuating electricity supply to the load which, in turn, may also fluctuate.

From this discussion, it will be clear that optimization of the entire RAPS system – solar array, storage battery, electricity-consuming devices – is critical for the facility to be commercially competitive. This optimization has to take into account factors such as latitude, meteorological conditions, solar cell efficiency, the use of static or steerable arrays, battery size, electrical losses, component life, and load profile. The development of RAPS systems which utilize solar-generated electricity is a highly specialized subject, and one in which the solar engineer needs to work closely with the battery expert.

12.2 BATTERIES IN SPACE

Earth-orbiting satellites and deep-space probes all derive their energy from the sun; photovoltaic panels are used to generate electricity. This is another, specialized application of solar power. In unmanned satellites, electrical power is required for communications, on-board computers, cameras and other surveillance equipment and, especially, for telemetry of data back to earth. Manned spacecraft have further requirements in regard to life-support services, and usually there is a fuel-cell system on board to supply the extra power required. As with terrestrial applications of photovoltaic electricity, a battery is used to store energy for periods of peak demand. It is also essential for the battery to provide energy when the spacecraft is in shadow and the solar cells are not functioning. Loss of communication at such times would be disastrous.

The basic specifications for a spacecraft battery are simple to define, if not to meet. The essential parameters are as follows.

- Guaranteed total reliability: if the battery fails, the satellite is virtually useless and an extremely valuable facility is lost.
- High specific energy: the battery and the associated power subsystem constitute a significant fraction of the mass of the satellite, and this mass has to be launched into orbit; any saving in battery mass would translate to an increase in payload, or a reduction in launch fuel.
- High energy density: room in the equipment bay is restricted, thus the smaller the battery the better.
- Long calendar-life and long cycle-life: the latter is particularly important when the satellite is in a low earth orbit.
- High energy efficiency: wastage of photovoltaic electricity is minimized.
- Ability to operate well in vacuum: this implies minimal heat production and good thermal contact with a heat sink, so that the battery does not overheat.
- Robust design: the battery must be engineered to withstand the considerable mechanical stresses experienced during rocket launch.

Some other factors which are often important in a battery specification are less so here. For example, the temperature range over which the battery is called upon to operate is less demanding, typically -10 to $+30°C$. Cost is also a secondary consideration – the battery is a vital component in a multi-million dollar vehicle. This latter factor is perhaps fortunate, as the development costs of producing a battery to meet the above specifications have been large and the market for the two categories of satellite (geosynchronous and low earth orbit) is quite small.

Geosynchronous satellites. Communications satellites used for relaying telephone, radio and television signals around the curved surface of the earth are in geosynchronous orbit, *i.e.* their orbit is synchronous with that of the earth. This means that the period of rotation of the satellites is just 24 h and they hold their position with respect to the earth beneath. In order for this to be so, the satellites must be situated above the equator at a height of 35 900 km, with an orbit which lies in the equatorial plane. With several such stationary satellites spaced around the equatorial orbit, world-wide communications coverage is possible. At the required height, the satellite is outside the shadow of the earth, as cast by the sun, and only moves into darkness during periods of eclipse. These periods are relatively rare, *i.e.* not more than a few times per year at most, and are of short duration, *i.e.* up to an hour. As the battery is normally only required during intervals of darkness to maintain communications, this is quite a light duty cycle. For a battery to be acceptable, the performance specification calls for 2000 cycles (charge and discharge at $C_2/2$ rate) to 70% DoD. In service use, the requirement will in fact be less severe, although the battery will be expected to perform satisfactorily for at least 10 years.

Low earth orbit satellites. These satellites are employed primarily for surveillance, *e.g.* in weather forecasting and in military defence, and orbit the earth at a variable height ($\sim$ 200 km) above the surface. At this height, they fall within the shadow of the earth and thus pass through periods of sunshine and darkness. The battery is charged in sunshine and discharged in darkness. Typically, the satellite makes 16 orbits per day and, accordingly, the battery undergoes the same number of charge–discharge cycles. If the satellite is to remain operational for five years, then the battery will need to withstand about 30 000 cycles. This is an enormously demanding specification, never before placed upon a battery, especially when combined with total reliability from failure.

The duty requirements of the above two categories of satellite illustrate the great extent to which the battery specifications may differ for two applications which, superficially, are very similar. Traditionally, custom-designed nickel–cadmium batteries were used in satellites. To provide acceptable cycle-life, the depth-of-discharge was often restricted to around 20% of the nominal (name-plate) capacity of the battery. This reduced the effective specific energy from the nominal value of 30 to 40 Wh kg^{-1} to less than 10 Wh kg^{-1}. The mass implications are obvious when the battery typically has to store about 1 kWh of energy. At about 100 kg, the battery becomes a major mass component of the satellite.

During the 1970s and 1980s, the nickel–hydrogen battery (see Section 11.4, Chapter 11) was developed specifically for use in spacecraft. In such service, the battery demonstrates several advantages over its nickel–cadmium counterpart, namely higher specific energy (50 to 60 Wh kg^{-1} compared with 30 to 40 Wh kg^{-1}); longer cycle-life at a given depth-of-discharge; an in-built means to monitor the state-of-charge, *i.e.* measurement of hydrogen pressure with a transducer. By the 1990s, nickel–hydrogen cells had largely replaced nickel–cadmium for use in satellites. In geosynchronous satellites, nickel–hydrogen batteries meet the performance specifications comfortably. In low earth orbit satellites, however, it is still necessary to use a shallow discharge to provide the very long cycle-life which is demanded; but even so, there are mass advantages compared with nickel–cadmium. It is interesting to speculate whether rechargeable nickel–metal-hydride or lithium-ion batteries will be used in satellites.

Space agencies are showing great interest in evaluating the lithium-ion

Table 12.1 *Specific energy and energy density of candidate space batteries*

Chemistry	Wh kg^{-1}		Wh l^{-1}	
	Cell	*Battery*	*Cell*	*Battery*
Nickel–cadmium	35	30	100	95
Nickel–hydrogen	56	46	50	40
Li-ion	120	110	250	180

system for both types of satellite. The much higher voltage of the cells, compared with rechargeable alkaline cells, provides an immediate boost in stored energy, while the low mass of lithium is advantageous for the specific energy. The gravimetric and volumetric energy densities of three contending space batteries, at both the cell and battery levels, are given in Table 12.1. The massive advantages which would accompany the use of lithium-ion batteries are abundantly clear. The European Space Agency plans to flight test lithium-ion space batteries.

12.3 STORAGE IN ELECTRICITY SUPPLY NETWORKS

The demand for medium/large industrial batteries falls into four broad market sectors:

- uninterruptible power supplies (UPSs) – discussed in Chapter 1 (Section 1.2) and Chapter 8 (Section 8.5);
- electric traction – discussed in the next section;
- smoothing of supply to demand for renewable energy sources – discussed in Section 12.1 for solar energy; rather similar considerations apply to other renewables (wind energy, wave energy) which are also erratic or periodic;
- storage of electricity in the utility supply system, which is considered here.

The demand for electricity fluctuates widely from summer to winter, from working day to holiday, and throughout the day from hour to hour and even minute to minute. The problem facing the utilities is to meet these fluctuations in demand in the most cost-effective manner. The seasonal fluctuations vary from country to country. In northern Europe, the peak demand is in winter for lighting (long hours of darkness) and heating. In warmer countries, especially southern Europe and southern USA, the peak demand is in summer for air-conditioning. At present, there is very little that can be done about seasonal storage, except for the possible introduction of pumped-hydro storage in particularly favourable locations. In the long-term, the possibility of using water electrolysis to form hydrogen has been debated. The gas can be stored in underground caverns and fed to fuel cells to re-generate electricity when required. To date, it has been concluded that this strategy is neither economic, nor very practicable. Batteries have nothing to offer for the seasonal storage of electricity.

As regards weekly and daily fluctuations in demand, Figure 12.2 shows a typical weekly load profile of a utility in the USA that is operating either with or without energy storage. As illustrated by the upper profile, the conventional supply is divided into four areas, *viz.* baseload (about 40% of installed capacity), intermediate (35%), peaking (11%), and reserve (the remaining 14%). Although the detail may change, this type of load variation applies to most other countries in which cheap off-peak electricity rates exist. Two facts are immediately evident:

- There is a marked diurnal pattern of demand during the working week, *i.e.* the demand falls from about 11 p.m., remains at a low level until about 6 a.m., and then rises during the day to high levels in the afternoon and the evening; a similar pattern is found in most industrialized countries.
- There is significantly less demand on Saturdays, and especially Sundays, when industrial usage is low.

The solid areas of the demand profile in Figure 12.2 (lower curve) show the unused baseload which is available, if stored, to meet the peak load (hatched areas of the profile). The baseload is supplied by a large (1 to 2 GW) fossil-fuelled or nuclear plant which is best kept running for long periods at a time and has the lowest generating costs. Intermediate demand is usually met by older, smaller plants with higher generating costs, while peak demand is furnished by reserve gas turbines or oil-fired plants at still higher costs. By storing the excess baseload electricity which is available over the week, it would be possible, in this example, to reduce the peak demand from 86% to around 75% of the installed capacity. Moreover, the higher baseload level (52% *vs.* 40%) would replace part of the intermediate generation. This is the concept of 'peak-shaving'. Such a strategy will enable central power stations to match system generation with changes in demand ('load-following') and, thereby, achieve better thermal efficiency, conserve fuel resources, and lower maintenance costs. Peak-shaving may also extend the useful life of the plant and mitigate air-quality problems. (Note, the term 'peak-shaving' is often confused with the term 'load-levelling'; the latter describes the more extensive use

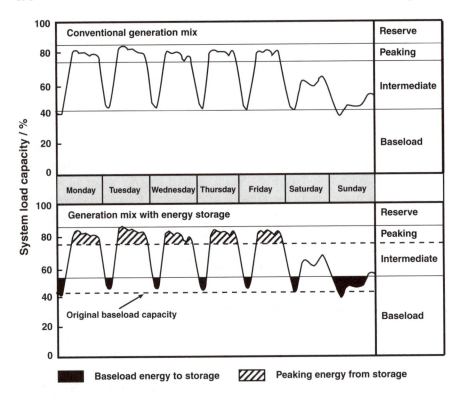

Figure 12.2   *Typical weekly load curve of an electricity utility in the USA*

of storage to eliminate most, or all, of the conventional intermediate generation.)

As noted in Chapter 1, several types of energy-storage systems are under investigation. These may be classified according to the form in which the energy is stored, namely potential energy (pumped hydro, compressed air); kinetic energy (flywheels); thermal energy (storage heaters, molten salts); chemical energy (batteries, methanol, hydrogen); electromagnetic energy (superconducting magnetic energy storage, SMES); electrostatic energy (capacitors, supercapacitors). A summary of the characteristics of some energy-storage systems is given in Table 12.2. Each technology has certain limitations. For example, pumped hydro and compressed air are location specific and cannot provide instantaneous power. The feasibility of using flywheels looks favourable on paper, but has yet to be proven in a full-scale demonstration programme. Thermal storage is thermodynamically inefficient, bulky, and subject to serious materials problems. SMES is extremely costly and is still very much at the research and development stage. Supercapacitors are an

Table 12.2 *Comparison of energy-storage systems*

	Pumped hydro	Compressed air	Flywheels	SMES	Batteries
Efficiency	~ 75%	~ 70% + fuel	~ 90%	~ 95%	~ 75%
Maximum energy	10 GWh	5 GWh	5 MWh	1.5 GWh	50 MWh
Maximum power	3 GW	1 GW	10 MW	1 GW	100 MW
Modular	No	No	Yes	Possible	Yes
Cycle-life	10 000	10 000	10 000	10 000	2000
Charge time	Hours	Hours	Minutes–hours	Minutes–hours	Hours
Siting ease	Poor	Poor	Good	Poor	Moderate
Lead time	Years	Years	Weeks	Years	Months
Environmental impact	Large	Large	Benign	Moderate	Moderate
Risk	Moderate	Moderate	Small	Moderate	Moderate
Thermal requirement	None	Cooling	Liquid nitrogen	Liquid helium	Air-conditioning
Maturity	Mature	Available	Embryonic	Embryonic	Mature

option only for power-quality applications in which the energy requirement is not large. Overall, these limitations suggest that batteries are presently the most attractive option. The financial incentive for adopting battery storage is indeed considerable, but everything hinges on the price at which large battery banks can be constructed and installed, as well as the performance and operational life of the batteries themselves.

Superimposed on the daily demand curve are short-term peaks associated with the personal habits of consumers, *e.g.* times of cooking meals, watching television, retiring to bed. Major demand 'blips' can arise quite suddenly, for example at the end of a popular television programme when viewers switch on electric kettles to make cups of tea or coffee. Knowledgeable power-station controllers can sometimes predict these incidents by studying television programme schedules. The response time to meet such sudden demands is of the order of seconds to minutes, if frequency adjustment or voltage reduction is to be avoided. Clearly, the requirement for such rapid service will not allow the start-up of additional generating units; the demand must be met from the 'spinning reserve'. (Note, this reserve also meets power emergencies which may arise from the unscheduled failure of generating units and/or transmission lines.) Batteries respond instantaneously to load changes and are therefore considered to be the ideal strategy for coping with short-term demands. The use of batteries would reduce or eliminate the need for spinning reserve and/or frequency adjustment.

Another important feature of electricity storage is that it is possible to load-level not only the generating plant but also the transmission and distribution system. Utilities are considering a move away from large, centralized, power plants towards smaller generation sources and storage units that are distributed geographically in close proximity to end-users (Figure 12.3). Such decentralization would provide many advantages, namely protection from major mishaps; decreased costs of transmission and distribution; lower power losses; improved local reliability and quality of supply; greater flexibility in matching capacity additions and retirements with changes in demand; reduced concerns over possible health effects of electric and magnetic fields; shorter construction times; less disruption from digging up roads and laying cables. By virtue of their demonstrated attributes of small size, modular construction, silent operation, negligible emissions, high efficiency and instantaneous response, battery energy-storage units can be placed practically on the distribution system at critical points in the network. Furthermore, batteries can be installed in incremental capacities which meet more closely the demand at the time. Distributed strategically in a utility network, a group of battery units, each sized between 1 and 10 MW, can yield greater benefits than a single, centrally located unit of equivalent total capacity. A convenient location for decentralized storage might be at one of the step-down transformers, with the most local of all being at the district transformer.

What battery specifications are set by the electricity utilities? The

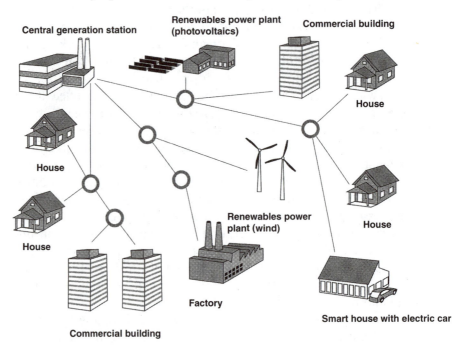

Figure 12.3   *Tomorrow's distributed utility*

prime consideration is the economics – what is the value of storage to the utility per kWh and per kW, and can battery manufacturers meet this target? For weekly or diurnal storage, the rates of charge and discharge are fairly conventional ($\sim$ hours), and so the focus is on energy stored (kWh). By contrast, the short-term peaks require rapid battery discharge, and perhaps rapid charge too. The emphasis here is on the power output of the battery (kW) rather than on the stored energy. In both types of service, the overall energy efficiency of the battery system (Wh output : Wh input) is crucial to the economics.

Until now, where battery storage has been demonstrated it has invariably been in the form of lead–acid batteries. This choice has been made on the grounds of availability and comparative cheapness. Also, these batteries provide high power output. Battery mass is not a major consideration for stationary batteries, provided they are floor mounted and not stacked. There are, however, certain drawbacks with the use of lead–acid batteries. Ideally, the energy-storage system will be situated outdoors under a simple, inexpensive shelter. Lead–acid batteries, however, require a certain level of maintenance (particularly, flooded-electrolyte types) and have to be housed in a dedicated room under temperature-controlled conditions. This room will have a considerable floor-area as the batteries cannot readily be stacked. Further, the cycle-life of lead–acid batteries under the conditions of use is barely adequate and their cost, when added to the other capital costs of the system (battery housing, power-conditioning equipment), is too high for many potential applications in the electricity supply industry. The utilities consider the ideal battery to be one of large size and low cost that can be stacked outdoors, requires no maintenance, is not sensitive to ambient temperature, and has a long life in terms of both years and charge–discharge cycles.

For centralized storage at the power station, the size of the facility required is more appropriate to an electrochemical engineering plant than to an assembly of unit cells. This is the reason why some utilities have shown an interest in flow batteries and redox batteries which are constructed on a plant basis, with the reactant chemicals housed in large reservoirs external to the cells. The demonstration Regenesys™ battery planned in the UK by Innogy plc (see Section 11.3, Chapter 11) will have a capacity of 100 MWh. If this system proves both technically successful and economically feasible, larger plants will undoubtedly be built. An artist's impression of the proposed Regenesys™ plant is shown in Figure 12.4. The technology is modular, comparatively easy to site, and separates the power rating (determined by the number and size of the modules) from the energy-storage capacity (determined by the size of the storage tanks). These features provide great flexibility in utility operations. Power outputs could range from 5 to 500 MW, and storage times from seconds to more than twelve hours. The system has been designed for a 20-year life. In principle, the smaller Regenesys™ units could be used for regional electricity storage.

12.4 ELECTRIC VEHICLES

Although the first experiments with electric traction were carried out in Scotland as long ago as 1837, it was not until the 1890s that the development of the lead–acid secondary battery was sufficiently advanced for the electric vehicle to become a practical reality. By the end of the 19th century, with the mass production of these rechargeable batteries, electric vehicles were a common sight on the roads of industrialized nations. Around 1903, the vehicle fleets in major cities such as London, New York and Paris each consisted of roughly equal numbers of internal-combustion-engined cars (automobiles), steam cars and electric cars. For a while, it was not obvious which of these three technologies was going to succeed in the race for powered transport. Gradually, however, it became apparent that the petrol-driven vehicle had the edge over its competitors in terms of both range and convenience of use. This view was strengthened by the invention of the electric self-starter by Charles Kettering in 1911. By removing the need to crank automobiles by hand it has been concluded that this device heralded the demise of the electric car. Ironically, it was the lead–acid battery which made this invention possible.

The Proceedings of the Institution of Civil Engineers (London) for 1903 contain an interesting paper on 'Electric Automobiles' that discusses the design and construction of the electric vehicles of the day and the batteries which propelled them, as well as the relative economics of

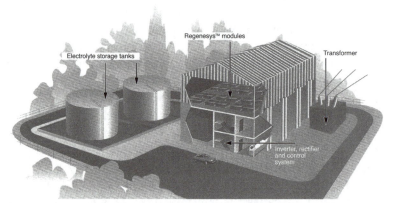

Figure 12.4   *Artist's impression of a 100 MWh Regenesys^{TM} energy-storage plant*
(By courtesy of Innogy plc)

petrol, steam and electric cars. The paper was presented to the Institution and prompted several interesting remarks, made in the subsequent discussion, which reflect on the technical and social history of the time. Some of the remarks were as follows.

The use of electricity for driving electric automobiles might expand so greatly that it would double, or even quadruple, the load of lighting-stations and would have a most important effect on their future.
(Authors' note: lighting stations were the power stations of the day.)

The subject under discussion was the use of electric automobiles in towns, as a means of reducing the number of horses and the consequent manure, slush and mud in the streets, which attended the use of horse-haulage.

Public service vehicles, private carriages, tradesmen's carts and wagons could all be driven by electricity at a price considerably below that for horseflesh and, in addition, would offer very considerable sanitary advantages, cleanliness and extra speed.

The cost of running electric vehicles will compare favourably with that for any other kind of self-propelled carriage and renders electric propulsion on roads, with its attendant advantages, more promising than any other system of motor carriage.

Colonel Crompton observed that '*he considered electric automobiles had a very good chance, especially under such conditions as would obtain in the case of a private family. Supposing, for instance, that the wife desired to do an afternoon's calling, he thought it could be done in an accumulator motor car at a cheaper rate than the fifteen shillings an afternoon charged for a carriage. The car could be driven by the ladies themselves, thereby saving the driver's wages. There was no doubt that ladies could drive accumulator cars; they only had to turn a handle, ring a bell, and start and stop the car and there was nothing to get out of order: whereas, on the other hand, it could not be said that, as a rule, ladies could be trusted with either petrol cars or steam cars.*'

In comparing electric cars and petrol cars the main point was the ease with which the speed could be varied in the electric car. In a petrol car the variation of speed was a mechanical problem which had not yet been satisfactorily solved; whereas in the electrical car it had been worked out and the speed could be varied with the greatest ease within wide limits. In the petrol car different sets of wheels had to be put into gear, and often the edges of the teeth were destroyed. It was necessary to get the speed right before changing the gear, otherwise a breakage occurred. Would it not be worthwhile to have an engine running continuously and driving a dynamo, and to have two motors to drive the wheels, thus doing away with the gearing and the incessant trouble caused by it?

This commentary illustrates: (i) the appalling sanitary conditions in city streets at the time; (ii) the embryonic state of the mechanical drive-train and the difficulty of controlling petrol vehicles; (iii) that technology forecasting is a most inexact science; (iv) how male chauvinism was rampant in 1903! It is also of interest to note that the last quotation may be one of the earliest references to an hybrid electric drive-train.

Within a few years of these optimistic statements, the electric road vehicle was essentially dead. What went wrong? Basically, following Kettering's invention of the self-starter, three further factors contributed to the passing of electric road traction:

- the rapid technological development of competing internal-combustion-engined vehicles;
- the evolution of the petroleum-refining industry and the setting up of an infrastructure of refuelling stations;
- the failure to develop a commercially viable traction battery which would give the vehicle adequate range and speed to satisfy the customer, coupled with failure to establish a network of recharging points.

These factors illustrate the importance of not under-rating the scope for advances in competing technologies, and also the need to develop a technology, such as electric traction, not in isolation, but as part of an overall system. In the case of electric road vehicles, the system is partly physical (the provision of an infrastructure of recharging points, repair shops, *etc.*) and partly socio-political (road transport policy, taxation policy, *etc.*).

By far the majority of electric vehicles manufactured during the 20th century were small, off-road vehicles, *e.g.* industrial trucks (platform trucks, fork-lift trucks), airport vehicles, locomotives for use in mines, golf carts, and wheelchairs. Electric vehicles were mostly chosen for their quietness and for their lack of pollution, especially for indoor applications. Road vehicles were limited to a few special situations, *e.g.* milk delivery trucks in the UK.

Renewed interest in electric road transport arose in the mid-1970s as concern heightened over the full realization that supplies of oil were indeed finite, and over the concentration of oil reserves in just a few countries. It was soon recognized that the prime requirement was for an advanced battery of higher specific energy than the lead–acid battery, so as to provide greater vehicle range between recharges. Later, in the mid-1980s, the concerns over petroleum supplies diminished as more reserves were found and political fears were not realized. In their place, a new anxiety arose over environmental pollution caused by automobiles, particularly in urban locations, and the general problem of global warming, thought to be due, in part, to build-up of carbon dioxide in the stratosphere.

To combat global, regional and local atmospheric pollution which emanates from road vehicles, the automotive industry started to design new electric cars and trucks, and to manufacture them in small numbers for trial and demonstration purposes. The major motor manufacturers in France, Japan and the USA all produced electric versions of one or more of their standard automobiles, while General Motors Corporation developed a completely new electric car, the *EV1* (*see Figure* 4.6, *Chapter* 4). *Meanwhile, battery companies and other laboratories were busy trying to develop new traction batteries, as described in Chapters 8 to 11.*

In the USA, a consortium of automotive companies, in partnership with the government, set up the United States Advanced Battery Consortium (USABC) with the goal of developing a high-performance battery suitable for traction purposes (see Section 10.5, Chapter 10). To date, there have been substantial research and development programmes on nickel–metal-hydride, lithium-ion and lithium–polymer batteries with the target of producing full-size EV batteries for demonstration purposes.

The USABC initiative stemmed from Californian legislation on environmental pollution which was adopted in 1990. Under the regulations, the largest automobile manufacturers (DaimlerChrysler, Ford, General Motors, Honda, Nissan, Toyota) were required to produce zero-emission vehicles (ZEVs). In model years 1998 through 2000, 2% of all new vehicles offered for sale in California by these manufacturers were to be ZEVs, and this percentage was to increase to 5% in model years 2001 and 2002, and 10% in model year 2003 and beyond. In 1996, the regulations were modified to allow additional time for the technology to develop. The requirement for 10% ZEVs in model year 2003 and beyond was maintained, but the sales requirement for model years 1998 through 2002 was eliminated. A further revision, in 1998, provided additional flexibility in the ZEV programme by allowing additional types of vehicle to be used to meet the legislative requirements. Under the 1998 amendments, the manufacturers must have 4% of their sales in model year 2003 classified as 'full' ZEVs; at present vehicle-production rates (i.e. model year 2000), this would equate to about

22 000 ZEVs. The remaining 6% of sales can be made up of extremely clean, advanced-technology vehicles, which are referred to as 'partial' ZEVs. This partial-allowance approach towards satisfying the ZEV requirement is intended to promote continued development of battery-powered electric and zero-emitting fuel-cell vehicles, while encouraging the development of other vehicles which have the potential for producing extremely low emissions. Over 2000 electric vehicles have already appeared on the roads of California (Table 12.3).

The USABC also established targets which a new traction battery would have to meet for an electric car to be commercially viable in the US environment. These targets are listed in Table 12.4. The mid-term targets were set as a step on the road to commercialization, in recognition of the difficulties faced by the vehicle developers. The long-term targets were thought to be necessary for full commercial realization of fleets of electric vehicles without any public subsidy. All the targets relate to batteries for use in family-size electric cars. Lesser goals may be acceptable by purchasers of small vehicles for purely urban use. Considerable progress has been made, and at least three batteries have come close to meeting most, though not all, of the mid-term targets. These are nickel–metal-hydride, lithium-ion and sodium–nickel-chloride. (The last-mentioned battery was a European development and not part of the USABC programme.) Cost is likely to be a problem especially in achieving the long-term target of < US$100 per kWh. Also, it must be borne in mind, as stated earlier, that the various targets are interactive and the demonstration that one

Table 12.3 *Electric vehicles deployed in Californian ZEV programme (to May 2000)*

Manufacturer	Model	Number	Battery type	City range[a] (km)	Highway range[b] (km)
DaimlerChrysler	EPIC	17	Lead–acid	113	105
	EPIC	93	Ni–MH[c]	148	156
Ford	Ranger	52	Lead–acid	135	111
	Ranger	327	Ni–MH	151	138
General Motors	EV1	606	Lead–acid	121	126
	EV1	162	Ni–MH	230	245
	S–10	110	Lead–acid	74	69
	S–10	117	Ni–MH	148	159
Honda	EV Plus	276	Ni–MH	201	169
Nissan	Altra	81	Li-ion	193	172
Toyota	RAV 4	486	Ni–MH	229	187

[a] US urban dynamometer driving schedule (UDDS).
[b] US highway fuel economy driving schedule (HFEDS).
[c] Nickel–metal-hydride.

Table 12.4 *USABC objectives for electric-vehicle battery packs*

USABC objective	Mid-term	Commercialization	Long-term
Primary objectives:			
• Specific energy (Wh kg^{-1}) ($C_3/3$ discharge rate)	80 (100 desired)	150	200
• Energy density (Wh l^{-1}) ($C_3/3$ discharge rate)	135	230	300
• Specific power (W kg^{-1}) (80% DoD per 30 s)	150 (200 desired)	300	400
• Specific power (regen.), 20% DoD / 10 s (W kg^{-1})	75	150	200
• Power density (W l^{-1})	250	460	600
• Recharge time (20 → 100% SoC) (h)	< 6	6 (4 desired)	3 to 6
• Fast recharge time, min	< 15 (40 → 80% SoC)	< 30 at 150 W kg^{-1} (20 → 70% SoC) (< 20 at 270 W kg^{-1} desired)	< 15 (40 → 80% SoC)
• End-of-life	20% degradation of power & capacity specification	20% degradation of power & capacity specification	20% degradation of power & capacity specification
• Calendar-life, years	5	10	10
• Life, cycles	600 at 80% DoD	1000 at 80% DoD 1600 at 50% DoD 2670 at 30% DoD	1000 at 80% DoD
• Life, urban miles	100 000	100 000	100 000
• Ultimate cost (US$/kWh)	< 150	< 150 (75 desired)	< 100
• Operating environment, °C	− 30 to + 65	20% loss at extremes of − 40 & + 50 (10% desired)	− 40 to + 85
• Continuous discharge in 1 h (no failure), % rated energy capacity	75	—	75
Secondary objectives:			
• Efficiency (%) ($C_3/3$ discharge rate, 6-h charge, at end-of-life)	75	80	80
• Off-tether pack energy loss — Self-discharge (%)	< 15 (in 48 h)	12 days: cumulative loss < 25% with some performance loss at extreme temperature limits	< 15 (in 1 month)
— Thermal loss	< 3.2 W / kWh (< 15% in 48 h)	3 days: < 15% with no performance loss	< 3.2 W / kWh (<15% in 48 h)
• Maintenance	zero	—	zero

can be met in isolation in the laboratory is no guarantee that, in conjunction with the other targets, it can be met in a vehicle trial. These considerations clearly show that traction battery development is slow, difficult and expensive.

In parallel with research into advanced batteries, other companies are working on fuel cells. These power sources are particularly well suited to larger vehicles, such as urban delivery trucks and buses. There are several types of fuel cell; discussion of their relative merits is outside the scope of this book. At present, the one system preferred for electric vehicles is the proton exchange membrane (PEM) fuel cell (also known as the solid polymer-electrolyte fuel cell, SPEFC).

Considerable interest is also being shown in the concept of the hybrid electric vehicle (HEV), which has two power sources. This may be an all-electric vehicle, with a fuel cell to provide range and a high-power battery to boost acceleration. Alternatively, it may be a heat-engine–battery hybrid, *i.e.* a partial ZEV, of which there are two basic types: the 'series HEV' and the 'parallel HEV', see Figure 12.5. In the series configuration, the output of a heat engine is converted into electrical energy through a generator which, either separately or jointly with a battery, powers a single drive-train. In one typical version, the series HEV would have a battery which is sufficiently large to meet the daily range and peak-power requirements for city driving, and a small heat engine (internal combustion engine or gas turbine) which is used to generate electricity purely as a 'range extender' for out-of-town driving. The battery is said to operate in the 'dual-power mode'. The series HEV is essentially an electric vehicle (EV) with an EV-sized battery and a small auxiliary engine.

By contrast, the parallel HEV has two distinct drive-trains such that the vehicle can be driven mechanically by a heat engine, or electrically by a battery–electric-motor, or by both. The heat engine is larger than that in a series HEV (but smaller than that in a conventional automobile) and is sized for steady highway driving. The independent battery system provides auxiliary power for acceleration and hill-climbing, accepts regenerative-braking energy, and restarts the engine in city traffic. In such duty, the battery has to furnish and absorb high, short bursts of current and is said to operate in the 'power-assist mode'. The parallel HEV corresponds to a conventional automobile with a smaller engine and a larger battery.

Whereas the series HEV is the simpler concept and has more scope for both lower fuel consumption and reduced emissions, the parallel HEV offers the easier option to manufacture as it can make use of existing technology for engines, gear boxes, induction motors, and lead–acid batteries. It should be noted that there can be a variety of different designs within each of the two broad categories of HEV, and that the engineering becomes even more complicated when supercapacitors (or even flywheels) are included. Heat-engine–battery hybrid vehicles are a complex subject. A full analysis of their development, relative advantages/disad-

Series hybrid

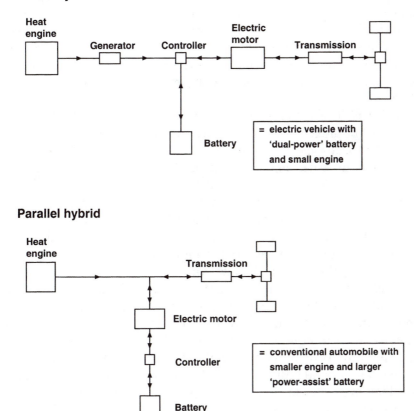

Parallel hybrid

Figure 12.5 *Series and parallel configurations of hybrid electric vehicles*

vantages and prospects is beyond the scope of this book. The major advantage of such vehicles is that the use of a second power source relaxes the battery specifications. Indeed, modern lead–acid batteries may well prove suitable, *e.g.* the CSIRO Double-Impact™ design (see Section 11.7, Chapter 11). The Toyota 'Prius' and Honda 'Insight' – parallel designs of hybrid car which, to date, have attracted the most attention – both use costly nickel–metal-hydride batteries. Clearly, the development and adoption of less-expensive lead–acid alternatives will increase the competitiveness of HEVs.

Leaving aside hybrid electric vehicles, which is the most promising traction battery for the pure-battery car? At present, there seem to be seven possible candidates whose principal attractions and limitations are summarized in Table 12.5. Based on a subjective judgement, a 'second division' of battery candidates might include zinc–bromine, sodium–sulfur, and mechanically rechargeable zinc–air. A survey of Table 12.5 shows that no one of the seven 'first division' batteries is entirely satisfactory for road traction applications. A few brief comments on each these batteries follow.

Lead–acid. This is the base technology against which all other candidates must be judged, since it is widely used in off-road traction applications. Most lead–acid traction batteries used to date in off-road duties have not been sealed and maintenance-free; these attributes apply to valve-regulated designs (see Section 8.7, Chapter 8) which are less capable of sustaining repetitive deep discharge. Major research programmes, particularly those funded by the Advanced Lead–Acid Battery Consortium (ALABC), are in place to solve this problem. The high mass of lead reduces dramatically the scope for improving the specific energy of lead–acid, and this is a principal weakness.

Nickel–cadmium. The manufacture of nickel–cadmium traction batteries is also a well-established industry. French developers of electric vehicles, in particular, have favoured this battery. For example, the battery is used in electric versions of small Citroën and Peugeot cars which are now available commercially (Figure 12.6). It has been argued that the high capital cost of these batteries may be justified by virtue of their long cycle-life, so that the cost per cycle is acceptable. A leasing arrangement would spread the cost over the life of the battery. Electricity is cheaper than petrol and the savings made would pay for the battery lease (so it is contended). The toxic nature of cadmium is still a matter of concern, as is the poor high-temperature performance of the battery.

Nickel–zinc. In many ways, nickel–zinc is the ideal compromise battery, except for its one major deficiency of short cycle-life. This appears to be a fundamental scientific problem. In recent years, research and development has resulted in a steady improvement in performance from ~ 100 to ~ 300 cycles, but there is still a way to go before the battery may be considered to be a viable candidate for traction applications.

Nickel–metal-hydride. Because of its success in the portable electronics market (*e.g.* mobile telephones), nickel–metal-hydride is seen as the ultimate rechargeable alkaline battery, the successor to nickel–cadmium and nickel–zinc. Several companies have built traction batteries, and field trials have been reported in a number of vehicles (Chrysler, Daimler-Benz, Fiat, Volvo, *etc.*). Nickel–metal-hydride batteries capable of meeting the power requirements of the USABC mid-term objectives have been demonstrated in more than 1000 vehicles in California (Table 12.3).

Table 12.5 *Principal attributes of candidate EV traction batteries*

Battery	Principal attractions	Limiting factors
Lead–acid (Chapter 8)	• Established industry • Lowest cost battery • Sealed (maintenance-free)	• Low specific energy • Moderate cycle-life (500) • Poor low-temperature performance
Nickel–cadmium (Chapter 9)	• Established industry • Long cycle-life ($\sim$ 2000)	• Moderate specific energy • Poor performance above 35°C • Toxic components • High cost
Nickel–zinc (Chapter 9)	• Reasonable specific energy	
Nickel–metal-hydride (Chapter 9)	• Good all-round candidate • Reasonable specific energy • High power • Long cycle-life • Tolerant to overcharge and overdischarge • Sealed (maintenance-free)	• Short cycle-life ($\sim$ 300) • High self-discharge • Poor charge-acceptance at high ambient temperature • High cost
Sodium–nickel-chloride (Chapter 11)	• High specific energy • Discharged-state assembly • Long cycle-life • No self-discharge • Successful fleet trials	• High operating temperature • Poor specific power at 80% DoD • Some heat loss • Uncertain cost
Lithium–ion (Chapter 10)	• High specific energy and power • Long cycle-life	• Fairly early stage of development • Accurate control of charge voltage needed • Safety? • High cost
Lithium–polymer (Chapter 10)	• Flexibility of configuration • High specific energy (anticipated)	• Very early stage of development

Present battery modules have specific energies of 65 to 70 Wh kg^{-1}, but it is considered that major increases are unlikely. The main drawback of this battery, however, is the high cost, which is even higher than that of nickel–cadmium.

Sodium–nickel-chloride. Extensive fleet trials of vehicles equipped with sodium–metal-chloride batteries have been conducted in Europe. These were highly successful in terms of performance, reliability and safety, and demonstrated that high-temperature batteries are a perfectly feasible concept. Remaining problems, now being addressed, are the limited power rating of the battery when near the fully-discharged state, and cost reduction. There is a fundamental question over whether the 'person in the street' (as opposed to the professional driver) can be trusted to keep the battery hot, by plugging in to the mains when the vehicle is parked for

Figure 12.6   *Electric versions of Citroën AX and Peugeot 106 cars with nickel–cadmium batteries*
(By courtesy of UK & International Press)

extensive periods. A further problem, unique to high-temperature batteries, is that it is not possible to first build small batteries for diverse applications before launching into the traction market. An electric version of the Mercedes 'A class' car equipped with sodium–nickel-chloride (ZEBRA) batteries is shown in Figure 12.7. With the subsequent amalgamation of Daimler-Benz and Chrysler, it was decided to abandon further work on this vehicle project.

Lithium-ion. The phenomenal success of this battery in portable electronics applications has encouraged several manufacturers to build full-sized traction modules. A battery developed by the Sony Corporation has been used in Altra vehicles produced by Nissan for the Californian ZEV initiative (Table 12.3). Using positive electrodes made from cobalt oxide, the design is the most expensive of all the candidate traction batteries, which explains why so much research is being devoted to the possibility of replacing cobalt in the oxide with manganese, a much cheaper and more widely available metal. Lithium-ion battery modules (2 to 3 kWh) based on manganese are presently under evaluation at the Akagi Testing Centre in Japan. The specific energy of the modules is $\sim 110 \, \text{Wh kg}^{-1}$. It should also be noted that there is some concern over possible safety issues associated with overcharge or other abuse of the lithium-ion system.

Lithium–polymer. The only reasons for including this system in the list of candidate traction batteries are its novel construction, exceptionally high specific energy and modest temperature of operation, together with the substantial research and development effort which is being undertaken in Canada, France, Japan and the USA to bring the technology to

Figure 12.7 *Electric version of Mercedes 'A class' car with sodium–nickel-chloride batteries*
(By courtesy of Beta Research & Development Ltd)

commercialization. Particular attention is being directed towards devising manufacturing techniques for unconventional, thin-film cells and batteries. Cycle-life is still a difficult issue, however, and it is considered by many that the prospects for a successful traction battery based on lithium metal are not good, even when using a solid polymer electrolyte. A compromise, much more likely to succeed, is a lithium-ion battery with a polymer membrane electrolyte. This is then both a lithium-ion and a lithium–polymer battery. The specific energy would be well below that of a battery using lithium metal, but could be greater than that of lithium-ion.

From this brief synopsis, it will be clear that no candidate possesses all the attributes sought in the ideal traction battery, and the jury is still out as to which battery will be preferred. It may well be that the answer will depend on the type of vehicle and on the battery manufacturer. The practical range provided by state-of-the-art batteries is considered less than desirable by most drivers. Curiously, though, after living with an electric vehicle, many owners have found that their actual driving patterns are less demanding than they had imagined. Automobile manufacturers also stress that the batteries will be too expensive for acceptable electric-vehicle costs. Major advances in technology and true mass-production are required to reduce battery costs substantially below present projections.

The above four applications – remote-area power supplies, satellites, utility energy-storage, electric vehicles – are presented in some detail to illustrate the many diverse considerations, both technical and non-technical, which determine the choice of a secondary battery for a particular duty. Throughout, we have emphasized the need for the design engineer to liaise closely with battery scientists and manufacturers so that the most suitable battery is selected for the intended application.

12.5   CONCLUDING REMARKS

In this book, we have outlined the principal characteristics of primary and secondary batteries which are in use today, or are under development. The aim has been to provide the reader with a useful introduction to the operation and the various applications of the batteries. The treatment is necessarily broad in content as, for each battery system, a thorough exposition would require a textbook of its own!

Throughout, an attempt has been made to relate the properties of the battery to the specification for the intended application. It is noted that the overall specification has to take account of environmental and safety aspects, as well as considerations of materials' availability and cost. Strong emphasis is placed on the importance of first defining the specification clearly, and then prioritizing the various criteria, before selecting a battery for a given duty. In this way, the limitations of the chosen battery will be clearly understood.

Many of the long-established batteries still enjoy a large share of the market. This is because their technology has advanced greatly in recent years to meet the ever-increasing demands of consumers for greater performance, reliability and life from batteries. In most cases it is only the basic electrochemical couple which is the same; the improvements have come from developments in materials science, technology and engineering. Intense research programmes have also resulted in keen competition from new battery chemistries. Two of these systems have now come to commercial fruition (nickel–metal-hydride, lithium-ion), while others such as lithium–polymer, sodium–metal-chloride and flow batteries do not lag far behind.

There is every reason to conclude that, in the decades ahead, the choice of batteries open to the consumer will be far greater than it has been in the past. The research work will continue to be driven by profound changes in society. Such changes will include:

- the proliferation of portable electronic and electrical devices;
- the introduction of electric and hybrid electric vehicles as a means to conserve oil supplies, to improve urban air quality, and to reduce greenhouse gas emissions;
- the expanding requirements for stationary energy storage – particu-

larly in the context of the decentralization of traditional electricity networks, the provision of power in remote areas, and the harnessing of renewable energy sources in efforts to achieve global energy sustainability.

Clearly, the future promises to be exciting for batteries.

Recommended Reading

BOOKS

Handbook of Battery Materials, ed. J. O. Besenhard, Wiley-VCH Verlag, Weinheim, Germany, 1999.

Lithium Ion Batteries: Fundamentals and Performance, ed. M. Wakihara and O. Yamamoto, Wiley-VCH Verlag, Weinheim, Germany, 1999.

D. A. J. Rand, R. Woods and R. M. Dell, *Batteries for Electric Vehicles*, Research Studies Press, Taunton, UK, 1998.

C. A. Vincent and B. Scrosati, *Modern Batteries: An Introduction to Electrochemical Power Sources*, 2nd edn., Arnold, London, 1997.

D. Berndt, *Maintenance-Free Batteries: Lead–Acid, Nickel/Cadmium, Nickel/Hydride. A Handbook of Battery Technology*, 2nd edn., Research Studies Press, Taunton, UK, 1997.

P. Reasbeck and J. G. Smith, *Batteries for Automotive Use*, Research Studies Press, Taunton, UK, 1997.

Handbook of Batteries, 2nd edn., ed. D. Linden, McGraw-Hill, New York, 1995.

Modern Battery Technology, ed. C. D. S. Tuck, Ellis Horwood, Chichester, 1991.

T. R. Crompton, *Battery Reference Book*, Butterworths, London, 1990.

Battery Technology Handbook, ed. H. A. Kiehne, Marcel Dekker, New York, 1989.

R. J. Brodd, *Batteries for Cordless Appliances*, Research Studies Press, Letchworth, UK, 1987.

A. Himy, *Silver–Zinc Battery: Phenomena and Design Principles*, Vantage Press, New York, 1986.

J. L. Sudworth and A. R. Tilley, *The Sodium–Sulphur Battery*, Chapman & Hall, London, 1985.

Solid State Batteries, ed. C. A. C. Sequeira and A. Hooper, Martinus Nijhoff, Dordrecht, 1985.

S. U. Falk and A. J. Salkind, *Alkaline Storage Batteries*, John Wiley & Sons, New York, 1969.

SCIENTIFIC JOURNALS AND PERIODICALS

Advanced Battery Technology, Seven Mountains Scientific, Inc., Boalsburg, PA 16827, USA.
E-mail: abt@7ms.com (Website: http://www.7ms.com)

Batteries International, Euromoney Publications PLC, London. E-mail: batteries–international@gwassoc.dircon.co.uk
(Website: http://www.batteriesint.com)

Electric & Hybrid Vehicle Technology, UK & International Press, Dorking, UK. E-mail: electric@ukintpress.com

Electric Vehicle Progress, Alternative Fuel Vehicle Group, New York.
E-mail: info@AltFuels.com (Website: http://www.AltFuels.com)

Journal of Power Sources, Elsevier Science Ireland Ltd., Shannon, Ireland. (Website: http://www.elsevier.nl)

USEFUL WEBSITES

Professional Associations

Advanced Lead–Acid Battery Consortium. http://www.alabc.org

Electric Vehicle Association of America. http://www.evaa.org

European Electric Road Vehicle Association. http://www.avere.org

International Power Sources Symposium (UK). http://www.ipss.org.uk

International Solar Energy Society. http://www.ises.org/ises.nsf

The British Wind Energy Association. http://www.bwea.com

The Electrochemical Society. http://www.electrochem.org/ecs

The UK Solar Energy Society. http://www.brookes.ac.uk/uk-ises

United States Advanced Battery Consortium.
http://www.ott.doe.gov/oaat/usabc.html

Battery Manufacturers and Developers

AEA Technology Batteries. http://www.aeatbat.com

Duracell, Inc. http://www.duracell.com

CEAC. http://www.ceac.com

Eagle–Picher Industries. http://www.eaglepicher.com

Electric Fuel Corporation. http://www.electric-fuel.com

Energizer. http://www.energizer.com

Evercel. http://www.evercel.com

Hawker. http://www.hawker.co.uk

Johnson Controls, Inc. http://www.johnsoncontrols.com

Ovonic Battery Company. http://www.ovonic.com/enstor.html

Panasonic USA. http://www.panasonic.com

Rayovac. http://www.rayovac.com

SAFT. http://www.saft.alcatel.com

Sanyo Corporation. http://www.sanyo.com

Sonnenschein GmbH. http://www.sonnenschein.org

Sony Corporation. http://www.sony.com

Ultralife Batteries, Inc. http://www.ulbi.com

Varta. http://www.varta.com

Subject Index

Accumulators, 2
Activation overpotential, 15, 17
Active material/mass, 12
Advanced batteries, 163 *et seq.*
Alkaline batteries,
  history, 5
  nomenclature, 97, 127
  rechargeable, 126 *et seq.*
Alkaline electrolytes, 126
Alkaline-manganese cells,
  primary, 21, 24, 28, 59–65
  rechargeable, 128–129
β-Alumina, 173–175
Ampere-hours, 18
Anode, 13
Anodic reaction, 13
Automotive batteries, 7, 13, 29, 103–112

Batteries,
  applications, vii, 5–8, 82–96, 188–214
  chargers, 35, 39–47
  charging, 1, 9, 34–46
  choice, 22 *et seq.*
  definitions, viii, 10 *et seq.*
  deterioration and failure, 98
  history, 1
  maintenance, 30
  markets, 5, 7, 8
  mass, 25
  medical, 82–86
  monitoring/management/control,
    47–51
  regulation, 42
  selection criteria, 32
  sizes, 8
  space, 193–196
  temperature, 27
  terminology, 10 *et seq.*
  uses, 6
  volume, 25
Battery (cycle) life, 31, 32, 97–98
Battery module, 11
Battery pack, 11
Beta-alumina, 173–175
Bipolar batteries, 93
Bobbin cells, 55
Button and coin cells, 64–69, 72–74

Capacitors,
  electrolytic, 1
  super and ultra, 183–187
Capacity of a cell, *see* Storage capacity
Cardiac pacemakers, *see* Batteries,
  medical
Cathode, 13
Cathodic reaction, 13
Cells,
  chemistry, 11,13, 98
  discharge and charge, 14–21
  equalizing, 39
  reversal, 39
  sizes, *see* Consumer cells
  voltage, 14, 23
Charge acceptance, 34
Charging of batteries, *see* Batteries,
  charging
Chemical energy, 1
Chemistry in cells, *see* Cells, chemistry
Combined heat and power (CHP), 1
Concentration overpotential, 15,17
Conductance meters, 49
Consumer cells/batteries, 6, 20, 23,
  54–79
Coulombic (in)efficiency, 29, 30

220

Current-collector, 12
Current density, 20
Cycle life, *see* Battery (cycle) life

Daniell, John, 2
Daniell cell, 2, 3
Defibrillators, *see* Batteries, medical
Depolarizer, 15
Depth of discharge, 19, 41
Discharge/charge curves, 9, 18–21
Dry cells, *see* Zinc/carbon cells

Edison, Thomas, 5
Electric door bell, 4
Electric vehicles, 176–177, 182–186, 202–214
Electrical efficiency, *see* Energy efficiency
Electricity generation and storage, 1, 196–202
Electrochemical cells, 10
Electrode plates, 10
Electrode potentials, 12–14
Electrode reaction, *see* Cells, chemistry
Electroforming, 4
Electrolysers, 1
Electrolysis, 1, 4
Electrolysis cells, 12
Electrolyte, definition, 10
Electroplating, 4
Electrorefining, 4
Energy density, 25
Energy efficiency, 29, 30
Energy storage, 1

Fauré, Camille, 5, 100
Flooded-electrolyte batteries, 12
Flywheels, 1
Formation process, 4
Fuel cells, 1, 208
Fused salts, *see* Molten salt electrolytes

Galvanic cells, 12
Gravimetric energy density, *see* Specific energy
Grid, *see* Current-collector

Half-cell reaction, 12
Hybrid electric vehicles, 185, 208–209
Hydrogen, 1

Inductance charging, *see* Batteries, charging
Intercalation electrodes, 143–145, 147–149
Internal resistance, 11
Internal short-circuit, 10, 43, 101
Iron–chromium cells, 169

Jellyroll cells, *see* Spirally-wound cells
Jungner, Waldemar, 5

Kinetic energy, 1

Lead–acid batteries, 100–125
   automotive, 7, 13, 29, 103–112
   Bolder™ design, 123–4
   charging, 1, 9, 34–36
   Cyclon™ design, 122–123
   discharge curves, 19
   Double Impact™ design, 185, 209
   electrodes, 13, 35, 105–110
   electrolytes, 102–103, 119
   flooded, 101
   forming reaction, 40–41
   history, 4, 5, 100–102
   leisure, 21, 40, 112, 113
   markets, 125
   mode of operation, 102–104
   pasted plate, 35, 100–101
   Planté cells, 4, 100, 114–117
   positive plate alloys, 105, 107, 111
   separators, 110
   stationary, 7, 102, 113–117
   submarine, 119–120
   tubular traction, 101, 117–120
   valve regulated/sealed, 101, 120–124
Leclanché, Georges, 2
Leclanché cells, *see* Zinc–carbon cells
Leisure batteries, 40, 41, 112–113
Lithium batteries,
   button and coin cells, 72–74
   cell chemistry, 71, 79–81
   cylindrical and prismatic cells, 74–79
   high temperature, 158–162
   primary, 70 *et seq.*
   rechargeable, 143 *et seq.*
Lithium electrode, 70
Lithium–iodine cells, 83–86
Lithium–ion batteries, 147–153